Lecture Notes in Computer Scien

T0238400

Commenced Publication in 1973
Founding and Former Series Editors:
Gerhard Goos, Juris Hartmanis, and Jan van Leeuwen

Daniel Zeng   Ivan Gotham
Ken Komatsu   Cecil Lynch
Mark Thurmond   David Madigan
Bill Lober   James Kvach
Hsinchun Chen (Eds.)

Intelligence and Security Informatics: Biosurveillance

Second NSF Workshop, BioSurveillance 2007
New Brunswick, NJ, USA, May 22, 2007
Proceedings

 Springer

Volume Editors

Daniel Zeng
University of Arizona, Tucson, USA, E-mail: zeng@eller.arizona.edu

Ivan Gotham
New York State Department of Health, Albany, USA, E-mail: ijg01@health.state.ny.us

Ken Komatsu
Arizona Department of Health Services, Phoenix, USA, E-mail: komatsk@azdhs.gov

Cecil Lynch
University of California at Davis, Sacramento, USA, E-mail: colynch@ucdavis.edu

Mark Thurmond
University of California, Davis, USA, E-mail: mcthurmond@ucdavis.edu

David Madigan
Rutgers University, New Brunswick, NJ, USA, E-mail: dmadigan@rci.rutgers.edu

Bill Lober
University of Washington, Seattle, USA, E-mail: lober@u.washington.edu

James Kvach
Lexington, VA, USA, E-mail: jandk30@aol.com

Hsinchun Chen
University of Arizona, Tucson, USA, E-mail: hchen@eller.arizona.edu

Library of Congress Control Number: 2007926312

CR Subject Classification (1998): H.4, H.3, C.2, H.2, J.3, I.6, G.3, K.4.1

LNCS Sublibrary: SL 3 – Information Systems and Application, incl. Internet/Web and HCI

ISSN   0302-9743
ISBN-10   3-540-72607-1 Springer Berlin Heidelberg New York
ISBN-13   978-3-540-72607-4 Springer Berlin Heidelberg New York

Springer is a part of Springer Science+Business Media

springer.com

© Springer-Verlag Berlin Heidelberg 2007

Typesetting: Camera-ready by author, data conversion by Scientific Publishing Services, Chennai, India
Printed on acid-free paper   SPIN: 12066490   06/3180   5 4 3 2 1 0

Preface

The 2007 NSF BioSurveillance Workshop (BioSurveillance 2007) was built on the success of the first NSF BioSurveillance Workshop, hosted by the University of Arizona's NSF BioPortal Center in March 2006. BioSurveillance 2007 brought together infectious disease informatics (IDI) researchers and practitioners to discuss selected topics directly relevant to data sharing and analysis for real-time animal and public health surveillance. These researchers and practitioners represented a wide range of backgrounds including but not limited to epidemiology, statistics, applied mathematics, information systems, computer science and machine learning/data mining.

BioSurveillance 2007 aimed to achieve the following objectives: (a) review and examine various real-time data sharing approaches for animal and public health surveillance from both technological and policy perspectives; (b) identify key technical challenges facing syndromic surveillance for both animal and human diseases, and discuss and compare related systems approaches and algorithms; and (c) provide a forum to bring together IDI researchers and practitioners to identify future research opportunities. We are pleased that we received many outstanding contributions from IDI research groups and practitioners from around the world. The one-day program included one invited presentation, 17 long papers, six short papers, and two posters.

BioSurveillance 2007 was jointly hosted by: the University of Arizona; University of California, Davis; Rutgers, The State University of New Jersey; and the University of Washington.

We wish to express our gratitude to all workshop Program Committee members and reviewers, who provided high-quality, valuable and constructive review comments. We would like to thank Ms. Catherine A. Larson and members of the Artificial Intelligence Laboratory and the Intelligent Systems and Decisions Laboratory at the University of Arizona for their excellent support. BioSurveillance 2007 was co-located with the 2007 IEEE International Conference on Intelligence and Security Informatics (ISI 2007). We wish to thank the ISI 2007 organizers and support staff for their cooperation and assistance. We also wish to acknowledge the Springer LNCS editorial and production staff for their professionalism and continued support for ISI and related events. Our sincere gratitude goes to all of the sponsors, especially the U.S. National Science Foundation as the main sponsor

May 2007

Daniel Zeng
Ivan Gotham
Ken Komatsu
Cecil Lynch
Mark Thurmond
David Madigan
Bill Lober
James Kvach
Hsinchun Chen

Organization

Organizing Committee

Conference Co-chairs

Hsinchun Chen	University of Arizona
James Kvach	(formerly with) Armed Forces Medical Intelligence Center, Department of Defense
Bill Lober	University of Washington
David Madigan	Rutgers University
Mark Thurmond	University of California, Davis

Program Co-chairs

Daniel Zeng	University of Arizona
Ivan Gotham	New York State Department of Health
Ken Komatsu	Arizona Department of Health Services
Cecil Lynch	University of California, Davis

Government Liaisons

Larry Brandt	National Science Foundation
Dale Nordenberg	Centers for Disease Control and Prevention
Henry Rolka	Centers for Disease Control and Prevention

Local Arrangements

Catherine A. Larson	University of Arizona

Program Committee

John Allegra	Emergency Medical Associates of New Jersey
Michael Ascher	Lawrence Livermore National Laboratory
John Berrezowski	Alberta Agriculture, Food and Rural Development, Canada
Judith Brillman	University of New Mexico
Ian Brooks	University of Illinois at Urbana-Champaign
David L. Buckeridge	McGill University
Howard Burkom	Johns Hopkins University
Kathleen Carley	Carnegie Mellon University

Wendy Chapman	University of Pittsburgh
Carlos Chavez-Castillo	Arizona State University
Dennis Cochrane	Emergency Medical Associates of New Jersey
Noshir Contractor	University of Illinois at Urbana-Champaign
Kevin Coonan	University of Utah and Health Data Security Inc.
Greg Cooper	University of Pittsburgh
Dan Desmond	SIMI Group
Millicent Eidson	New York State Department of Health
Daniel Ford	IBM Almaden Research
Carol Friedman	Columbia University
Sherrilynne Fuller	University of Washington
Colin Goodall	AT&T Labs
Richard Heffernan	NY City Department of Health and Mental Hygiene
Alan Hevner	University of South Florida
Paul Hu	University of Utah
Xiaohua (Tony) Hu	Drexel University
Jesse Huang	Peking Union Medical College, China
Jeffrey Johnson	San Diego County Health and Human Services Agency
James Kaufmann	IBM Almaden Research
Ken Kleinman	Harvard University
Eileen Koski	Quest Diagnostics
Yang Kuang	Arizona State University
Brian La Forgia	Bergen County Department of Health Services
Colleen Martin	Centers for Disease Control and Prevention
Marcello Pagano	Harvard University
Marc Paladini	New York City Department of Health and Mental Hygiene
Fred Roberts	Rutgers University
Debbie Travers	North Carolina Public Health Information Network
Jason T. L. Wang	New Jersey Institute of Technology
Xiaohui Zhang	Scientific Technologies Corporation

Table of Contents

Part I: Long Papers

Biosurveillance Data Feed and Processing

Early Outbreak Detection Using an Automated Data Feed of Test
Orders from a Veterinary Diagnostic Laboratory 1
*Loren Shaffer, Julie Funk, Päivi Rajala-Schultz, Garrick Wallstrom,
Thomas Wittum, Michael Wagner, and William Saville*

Chinese Chief Complaint Classification for Syndromic Surveillance 11
*Hsin-Min Lu, Chwan-Chuen King, Tsung-Shu Wu, Fuh-Yuan Shih,
Jin-Yi Hsiao, Daniel Zeng, and Hsinchun Chen*

Incorporating Geographical Contacts into Social Network Analysis for
Contact Tracing in Epidemiology: A Study on Taiwan SARS Data 23
*Yi-Da Chen, Chunju Tseng, Chwan-Chuen King,
Tsung-Shu Joseph Wu, and Hsinchun Chen*

Biosurveillance Models

A Model for Characterizing Annual Flu Cases 37
Miriam Nuño and Marcello Pagano

Population Dynamics in the Elderly: The Need for Age-Adjustment in
National BioSurveillance Systems 47
Steven A. Cohen and Elena N. Naumova

Outbreak Detection Algorithms

Data Classification for Selection of Temporal Alerting Methods for
Biosurveillance 59
Howard Burkom and Sean Murphy

High Performance Computing for Disease Surveillance 71
David Bauer, Brandon W. Higgs, and Mojdeh Mohtashemi

Towards Real Time Epidemiology: Data Assimilation, Modeling and
Anomaly Detection of Health Surveillance Data Streams 79
*Luís M.A. Bettencourt, Ruy M. Ribeiro, Gerardo Chowell,
Timothy Lant, and Carlos Castillo-Chavez*

Algorithm Combination for Improved Performance in Biosurveillance
Systems 91
Inbal Yahav and Galit Shmueli

Decoupling Temporal Aberration Detection Algorithms for Enhanced
Biosurveillance 103
      Sean Murphy and Howard Burkom

Multiple Data Streams

Assessing Seasonal Variation in Multisource Surveillance Data: Annual
Harmonic Regression 114
      Eric Lofgren, Nina Fefferman, Meena Doshi, and Elena N. Naumova

A Study into Detection of Bio-Events in Multiple Streams of
Surveillance Data 124
      Josep Roure, Artur Dubrawski, and Jeff Schneider

Informatics Infrastructure and Case Studies

A Web-Based System for Infectious Disease Data Integration and
Sharing: Evaluating Outcome, Task Performance Efficiency, User
Information Satisfaction, and Usability 134
      Paul Jen-Hwa Hu, Daniel Zeng, Hsinchun Chen,
      Catherine A. Larson, and Chunju Tseng

Public Health Affinity Domain: A Standards-Based Surveillance System
Solution 147
      Boaz Carmeli, Tzilla Eshel, Daniel Ford, Ohad Greenshpan,
      James Kaufman, Sarah Knoop, Roni Ram, and Sondra Renly

The Influenza Data Summary: A Prototype Application for Visualizing
National Influenza Activity.. 159
      Michelle N. Podgornik, Alicia Postema, Roseanne English,
      Kristin B. Uhde, Steve Bloom, Peter Hicks, Paul McMurray,
      John Copeland, Lynnette Brammer, William W. Thompson,
      Joseph S. Bresee, and Jerome I. Tokars

Global Foot-and-Mouth Disease Surveillance Using BioPortal 169
      Mark Thurmond, Andrés Perez, Chunju Tseng,
      Hsinchun Chen, and Daniel Zeng

Utilization of Predictive Mathematical Epidemiological Modeling in
Crisis Preparedness Exercises 180
      Colleen R. Burgess

Part II: Short Papers

Ambulatory e-Prescribing: Evaluating a Novel Surveillance Data
Source... 190
      David L. Buckeridge, Aman Verma, and Robyn Tamblyn

Detecting the Start of the Flu Season   196
  Sylvia Halász, Philip Brown, Colin R. Goodall, Arnold Lent,
  Dennis Cochrane, and John R. Allegra

Syndromic Surveillance for Early Detection of Nosocomial Outbreaks . . .   202
  Kiyoshi Kikuchi, Yasushi Ohkusa, Tamie Sugawara,
  Kiyosu Taniguchi, and Nobuhiko Okabe

A Bayesian Biosurveillance Method That Models Unknown Outbreak
Diseases   209
  Yanna Shen and Gregory F. Cooper

Spatial Epidemic Patterns Recognition Using Computer Algebra   216
  Doracelly Hincapié and Juan Ospina

Detecting Conserved RNA Secondary Structures in Viral Genomes:
The RADAR Approach   222
  Mugdha Khaladkar and Jason T.L. Wang

Part III: Extended Abstracts

Gemina: A Web-Based Epidemiology and Genomic Metadata System
Designed to Identify Infectious Agents   228
  Lynn M. Schriml, Aaron Gussman, Kathy Phillippy, Sam Angiuoli,
  Kumar Hari, Alan Goates, Ravi Jain, Tanja Davidsen,
  Anu Ganapathy, Elodie Ghedin, Steven Salzberg, Owen White, and
  Neil Hall

Internet APRS Data Utilization for Biosurveillance Applications   230
  Tanya Deller, Rochelle Black, Francess Uzowulu,
  Vernell Mitchell, and William Seffens

Author Index   233

Early Outbreak Detection Using an Automated Data Feed of Test Orders from a Veterinary Diagnostic Laboratory

Loren Shaffer[1], Julie Funk[2], Päivi Rajala-Schultz[1], Garrick Wallstrom[3], Thomas Wittum[1], Michael Wagner[3], and William Saville[1]

[1] Department of Veterinary Preventive Medicine, The Ohio State University, Columbus, Ohio 43210
[2] National Food Safety and Toxicology Center, Michigan State University, East Lansing, Michigan 44824
[3] Department of Biomedical Informatics, University of Pittsburgh, Pittsburgh, Pennsylvania 15219

shaffer.45@osu.edu, funkj@cvm.msu.edu, rajala-schultz.1@osu.edu, garrick@cbmi.pitt.edu, wittum.1@osu.edu, mmw@cbmi.pitt.edu, saville.4@osu.edu

Abstract. Disease surveillance in animals remains inadequate to detect outbreaks resulting from novel pathogens and potential bioweapons. Mostly relying on confirmed diagnoses, another shortcoming of these systems is their ability to detect outbreaks in a timely manner. We investigated the feasibility of using veterinary laboratory test orders in a prospective system to detect outbreaks of disease earlier compared to traditional reporting methods. IDEXX Laboratories, Inc. automatically transferred daily records of laboratory test orders submitted from veterinary providers in Ohio via a secure file transfer protocol. Test products were classified to appropriate syndromic category using their unique identifying number. Counts of each category by county were analyzed to identify unexpected increases using a cumulative sums method. The results indicated that disease events can be detected through the prospective analysis of laboratory test orders and may provide indications of similar disease events in humans before traditional disease reporting.

1 Introduction

Prompt detection of outbreaks might provide for earlier intervention efforts that result in minimizing their overall impact [11], [19], [28], [32]. Some animals are susceptible to infection from many of the same pathogens as humans, sometimes showing signs of disease earlier [1], [12]. Therefore, animals might be used as sentinels and provide for earlier recognition of disease outbreaks that could affect humans. As pet animals share much of the same environment as their human owners, they especially might prove to be valuable outbreak sentinels [2].

Most of the current disease surveillance systems used for animal populations are considered inadequate for detecting outbreaks of emerging disease, potential acts of bioterrorism, or outbreaks resulting from pathogens for which the system was not

D. Zeng et al. (Eds.): BioSurveillance 2007, LNCS 4506, pp. 1–10, 2007.
© Springer-Verlag Berlin Heidelberg 2007

specifically designed for in a timely manner [16], [20], [21], [24]. Such functionality in animal-based systems has been considered important to our overall bioterrorism and disease outbreak preparedness capabilities [13], [14], [25], [28], [31], [32], [34]. Syndromic surveillance methods utilize population health indicators to warn of potential outbreaks earlier than reports of confirmed diagnoses. Although many sources of data have been investigated for syndromic surveillance in humans, there is paucity in the literature describing similar studies in animals [17].

Laboratories are recognized as important sources of data for disease surveillance in animals as well as humans [9]. Test orders for specimens submitted to commercial medical laboratories have been utilized as one of the data sources for syndromic surveillance in humans [5], [33]. Most of the private veterinary practitioners in the United States also submit specimens to commercial laboratories for diagnostic testing [15]. Through the utilization of data from these commercial laboratories, we might possibly achieve the benefit of the aggregation of many veterinary providers across a wide geographic area. Such centralized aggregation of data may be important in detecting certain outbreaks [11]. The results of a previous investigation conducted by us demonstrated the representation of companion animals in select veterinary diagnostic laboratory (VDL) data and indicated the potential for identifying clusters of cases through analysis of the aggregated orders [27].

Although laboratory analyses are not as frequently a part of the veterinary care of pet animals compared to the medical care of humans [31], we hypothesize that the consistency of test orders over time is such that increases in cases of disease will result in detectable increases in the number of test orders submitted by veterinarians that can be identified using prospective analysis.

2   Methods

We conducted a prospective study of laboratory orders submitted to IDEXX Laboratories, Inc. (Westbrook, Maine) for specimens originating from veterinary clinics in Ohio between September 1, 2006 and November 30, 2006. IDEXX transferred once daily to a server located at the Real-time Outbreak and Disease Surveillance (RODS) Laboratory (University of Pittsburgh, Pennsylvania), via secure file transfer protocol, an automatically generated text file containing records of laboratory orders for specimens received within the previous 24-hour period. Each record included the accession number assigned by IDEXX to the specimen, date and time that IDEXX received the specimen, 5-digit ZIP code of the clinic submitting the specimen, species of animal, and numerical code/s of the laboratory product/s ordered.

2.1 Mapping Laboratory Orders to Syndromic Category

We distributed a list of product descriptions ordered during a 2-week period to ten small and large animal veterinarians asking them to consider the diseases that they might use each product to confirm or rule out during the diagnostic process. The veterinarians then assigned each product to syndromic categories based on the expected presentation of these diseases. Eight categories were considered initially: respiratory, gastrointestinal, neurologic, behavioral, dermal, reproductive, non-specific, and sudden death. Seven of the ten surveyed veterinarians returned the categorized lists (Table 1). The behavioral and sudden death categories were subsequently removed based on zero responses from the surveyed veterinarians for these categories.

In addition to the surveyed veterinarians, two IDEXX laboratorians also reviewed the list of products. Based on their input and advice, five categories were added to further describe many of those products that had been classified into the non-specific category. These additional categories were endocrine, hepatic, infectious, febrile, and renal. Records were mapped to syndromic category based on the identifying number for the laboratory product ordered and appropriately classified as the server received them.

2.2 Statistical Analysis

We used frequency analysis to describe the representation of species groups and distribution of accessions by day of the week. The percentage of the total daily records included in the dataset for each 24-hour period was used to describe availability of records.

2.3 Detection Method

A cumulative sums (CuSum) method was used to analyze category counts, as records were received, for each Ohio County, as determined by the ZIP code. The value of the CuSum was calculated as

$$S_t = \max\{0, S_{t-1} + (X_t - (\mu_t + 0.5\sigma_t))/\sigma_t\}. \tag{1}$$

where X_t was the observed count at time t, μ_t the expected count (baseline), and σ_t the standard deviation of the counts used to determine the baseline. Daily analysis was performed automatically using the count from the current and previous six days for the observed value. A moving 7-day period was chosen to reduce the anticipated day-of-week effect in the data. The expected value was calculated by averaging the weekly counts for the previous 4-week period. We defined alerts as instances when the CuSum value equaled or exceeded five.

An alert period was defined as at least two consecutive days where the CuSum value exceeded the threshold. By using this two-in-a-row rule we were able to somewhat reduce the impact of single-day increases on weekly counts. Using this rule has been shown to increase the robustness of CuSum methods [22]. Alerts were considered for all syndromic categories except non-specific, which was mostly comprised of general screening tests such as blood chemistries. We investigated alerts by identifying the specific laboratory product or products involved and contacting select veterinarians located in the same area as the alert asking about their impressions of disease activity. Veterinarians may or may not have been IDEXX clients.

Table 1. Syndrome category descriptions distributed to veterinarian sample for grouping laboratory products

Example Diseases	Clinical Presentation	Syndrome Category
Glanders, Bordetella, Aspergillosis	• Coughing • Dyspnea • Nasal discharge	Respiratory
Salmonellsis, Clostridia-associated enterocolitis, Campylobacter	• Diarrhea • Vomiting • Colic	Gastrointestinal
Heartwater, plant poisoning, Botulism, Tetanus	• Convulsions • Paralysis • Staggering • Disturbed vision	Neurologic
Poxvirus, allergies, Foot and Mouth Disease	• Abscesses • Rash • Hair loss • Vesiculation	Dermal
Brucellosis, chronic Leptospirosis	• Retained placenta • Abortion • Orchitis	Reproductive
Plague, Tularemia, Anemia, early Leptospirosis	• Lethargy • Malaise • Weakness • Fever without defining associated sign	Non-specific
acute swine erysipelas, Anthrax, Red Water Disease	• Rapid onset of death without apparent cause • Death occurring after brief display of illness	Sudden Death
Rabies, Listeriosis	• Change in appetite without defining associated signs • Unexplained aggression • Disorientation	Behavioral

3 Results

3.1 Data Transfer

During the pilot, the daily transfer of data from IDEXX Laboratories was interrupted twice. The first interruption began on September 7 and continued through September 28. This interruption in data transfer occurred because the workflow involved in the transfer had been unscheduled and the job was mistakenly shut down. The second interruption occurred October 6 through October 9 for unknown reasons. The interruptions affected the transfer of 10,847 (22.6%) records. IDEXX forwarded records that were created during these times of interruption once the data feed was re-established providing for a complete time-series.

The pilot system relied upon transfer of data from IDEXX that was being queued in a test environment. The reliability of this environment was knowingly not as stable as a production environment would be. The interruptions experienced during this pilot would not be expected in a more stable production platform.

3.2 Descriptive Statistics

During the study period, IDEXX transferred records for 48,086 accessions. Specimens originated throughout Ohio and appeared to correlate with the population of each area. Accessions displayed an obvious and predictable day-of-week effect (Figure 1) with Sundays, Mondays, and days following holidays representing days with the lowest volume. Species represented by the accessions included canine (70.1%), feline (25.6%), and equine (2.1%).

An important consideration for the designers of any syndromic surveillance system is the timely availability of data [6], [30]. Earlier detection being the overall goal, the system must receive records, with the appropriate information for analysis, within a period that provides for improved timeliness of detection compared to traditional reporting systems. Excluding the accessions that occurred during the interruption periods (n=10,847), on average, 95% of daily records were received with the next day's dataset (Figure 2). Almost all (99.4%) records were received by the fourth 24-hour period.

3.3 Aberration Detection

The system identified nine alert periods during the study period using the CuSum detection method as previously described. All of the alerts involved canines and/or felines. The number of accessions generating the alerts ranged from eight to 43. No cause could be determined for three of the nine (33.3%) alert periods and two (22.2%) were possibly related to breeding operations that existed in the area (e.g. screening of litters for pathogens). Two (22.2%) others were potentially the result of provider interest. One veterinary practice located in an area where a gastrointestinal alert occurred reported being especially interested in educating clients about the risks from parasite ova. Another provider in an area where an endocrine alert occurred had recently been ordering an increased number of thyroid tests that were unrelated to increases in clinical disease. The remaining two (22.2%) alert periods were linked to verified disease activity in the pet population during the time of the alert.

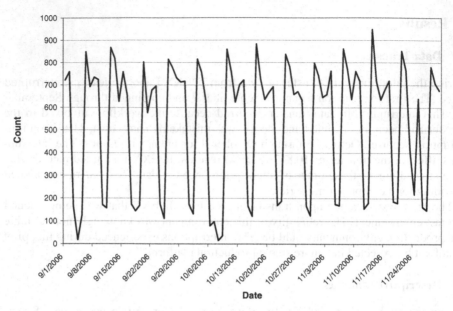

Fig. 1. Counts of specimens received by IDEXX from veterinary clinics in Ohio from September 1 through November 30, 2006

3.4 Case Reviews

On September 11, 2006, the system generated an alert in Preble County located in western Ohio. Cases (20 cats and 2 dogs) were equally distributed between two ZIP codes. Follow-up with area veterinarians confirmed that many small animal practices were treating an increased number of animals that lived or spent a significant amount time out-of-doors for unspecified gastrointestinal distress. Following consultation with the Ohio Department of Natural Resources, veterinarians suspected that the cases may have resulted from corona virus infections acquired from rodents (Melissa Howell, Preble County Health Department, personal communication). An increased number of rodents were noted in the area, coinciding with the harvesting of local grain fields. Veterinarians speculated that pets may have captured and consumed some of the rodents, resulting in the self-limiting intestinal condition. Although health authorities received no reports of human cases, the Real-time Outbreak and Disease Surveillance System used by the Ohio Department of Health indicated significant increases in both gastrointestinal-related chief complaints of emergency department patients and sales of anti-diarrheal medication in these areas during this time (L.S., unpublished data, 2006).

The pilot system generated a gastrointestinal alert for Lake County in northeastern Ohio on September 4, 2006. This alert included test orders for specimens originating from ten cats and three dogs submitted by clinics in two ZIP code areas. A local veterinarian from this county telephoned the State Public Health Veterinarian on

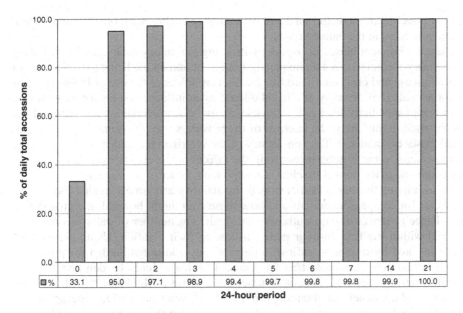

| □% | 33.1 | 95.0 | 97.1 | 98.9 | 99.4 | 99.7 | 99.8 | 99.8 | 99.9 | 100.0 |

Fig. 2. Delay in receipt of daily records from IDEXX during prospective pilot

September 26, 2006 to inquire about a number of clients that had brought their pets presenting with vomiting and diarrhea (Nancy Niehaus, Lake County Health Department, personal communication). These clients had shared with the local veterinarian that they also were experiencing diarrhea. The Lake County Health Department reported on October 4, 2006 that they were investigating "a cluster of diarrheal illness in humans and their associated pet dogs."

4 Discussion

The primary purpose of this study was to explore the feasibility of using prediagnostic data from a VDL in a prospective manner to detect unexpected increases in the number of disease cases that might indicate an outbreak. We evaluated the feasibility by first determining the stability of electronic records and the success of automatically transferring them from the VDL for analysis, measured in terms of the percentage of complete records received in a timely manner. We then considered the representation of the records both by species of animal and geographic distribution. Finally, we investigated the alerts generated by the pilot system to validate if they might be associated with increases of disease.

While no single data source provides the capability to detect all outbreaks that may occur, veterinary providers may be desirable sources to include in surveillance activities for bettering our capabilities of detecting those outbreaks that result from emerging pathogens and potential bioweapon agents [10], [12], [16], [25], [32]. The change in the number of laboratory orders submitted by veterinary providers may be a valuable proxy to measure the number of individual cases they are treating. An increase in the number of these individual cases may result from an outbreak, detection of which may be possible through the analysis of aggregated laboratory orders counts from several providers in the outbreak area.

There are inherent biases to consider with using laboratory data. Laboratory testing in veterinary medicine is not as frequently used as in human medicine [31]. Severity of clinical disease and cost benefit are two factors that influence the use of laboratory testing for animals [26]. Severity of clinical disease as an influence on testing may provide for increased specificity since only animals with true disease/condition are included. As demonstrated in this study, the interests of the providers may also contribute to the potential biases encountered. The consistency of the veterinarians' ordering behavior may help to control some bias by recognizing the effects in the counts over time and how they contribute to the normal baseline (i.e. expected number of test orders).

The results of this study demonstrated the stability and timely availability of test order data for companion animals and how those data might be used in a prospective surveillance to detect disease outbreaks. A significant number of daily records were received within the first 24-hour period following their creation. Using pre-existing data, generated by routine workflow, minimizes any additional burden for providers. Employing an automated data transfer protocol further reduces burden and is an essential benefit to support a sustained surveillance effort [16], [17], [31]. This system also achieved the benefit of obtaining provider-level data from a wide geographic area through a single source, creating no additional work for the veterinary providers and minimal work to establish and maintain the automated transfer mechanism for records from the VDL.

The results from this study also indicated that increases in the number of test orders submitted by veterinarians can be detected in a timely manner using prospective analysis. The development of the syndrome categories and the detection method used most likely influenced the alerts generated by this pilot system. We described two alerts that linked unexpected increases in test orders to increased incidence of disease. One of these alerts may also have provided warning of human cases of disease. The number of true and verifiable outbreaks of disease that occur limits determining the performance of an outbreak detection system [18], [29]. We lacked such a gold standard in this study. Therefore, we considered attempting any estimates of sensitivity, specificity, or positive predictive value to be inappropriate. Additional investigation, following refinement of the syndrome categories, might be beneficial for better evaluating the potential of such a system to detect outbreaks of disease.

The results support the continued consideration of VDL data by demonstrating the quality of data available, the ability to transfer and analyze the data in a timely manner, and the potential for detecting real disease events in the surveillance population. The true measure of a surveillance system lies in its usefulness [7]. Additional benefits from this method of surveillance may exist that contribute intangible value to the system [4], [8]. Previous studies found that regular reports of conditions were considered beneficial by data providers [3], [23]. While prospective analysis of orders includes methods designed to detect aberrant increases, reports of area syndromic trends may be valuable to veterinarians when treating individual animals as part of their practice. The addition of test results might also provide reports beneficial for veterinarians while potentially improving the specificity of outbreak detection. Input from all potential end users should be considered when further developing the utility of this type of surveillance system to ensure its maximum benefit.

Acknowledgments. The authors wish to thank Dr. Bill Wallen, Gary Watson, and Robert Ledford at IDEXX Laboratories, Inc. for granting access to use these data and for their technical advice and assistance in establishing the transfer mechanism.

References

1. Babin SM et al. Early detection of possible bioterrorist events using sentinel animals. The 131st Annual Meeting of the American Public Health Association. 2003.
2. Backer L et al. Pet dogs as sentinels for environmental contamination. Sci Total Environ 2001;274:161-9.
3. Bartlett PC et al. Development of a computerized dairy herd health data base for epidemiologic research. Prev Vet Med 1986;4:3-14.
4. Begier EM et al. The National Capitol Region's Emergency Department Syndromic Surveillance System: Do Chief Complaint and Discharge Diagnosis Yield Different Results? Emerg Infect Dis 2003;9:393-6.
5. Bradley CA et al. BioSense: Implementation of a National Early Event Detection and Situational Awareness System. MMWR Morb Mortal Wkly Rep 2005;54:11-9.
6. Buehler JW et al. Framework for Evaluating Public Health Surveillance Systems for Early Detection of Outbreaks. MMWR Recomm Rep 2004;53.
7. Buehler JW. Surveillance. In: Rothman KJ, Greenland S, eds. Modern Epidemiology. Philadelphia, PA: Lippincott Williams & Wilkins, 1998:435-57.
8. Buehler JW et al. Syndromic Surveillance and Bioterrorism-related Epidemics. Emerg Infect Dis 2003;9:1197-204.
9. Conner CF. Review of efforts to protect the agricultural sector and food supply from a deliberate attack with a biological agent, a toxin or a disease directed at crops and livestock. July 20, 2005. Bio-security and Agro-terrorism. 2005.
10. Conti L. Petborne Zoonoses: Detection and Surveillance Challenges. Burroughs, T., Knobler, S., and Lederberg, J. The Emergence of Zoonotic Diseases: Understanding the Impact on Animal and Human Health. 2002. Washington, DC, National Academy Press.
11. Dato V, Wagner MM, Fapohunda A. How Outbreaks of Infectious Disease are Detected: A Review of Surveillance Systems and Outbreaks. Public Health Rep 2004;119:464-71.
12. Davis RG. The ABCs of bioterrorism for veterinarians, focusing on Category A agents. J Am Vet Med Assoc 2004;224:1084-95.
13. Doherr MG, Audige L. Monitoring and surveillance for rare health-related events: a review from the veterinary perspective. Philos Trans R Soc Lond B Biol Sci 2001;356:1097-106.
14. Engle MJ. The Value of an "Early Warning" Surveillance System for Emerging Diseases. The Value of an "Early Warning" Surveillance System for Emerging Diseases. 2000. National Pork Board.
15. Glickman LT et al. Purdue University-Banfield National Companion Animal Surveillance Program for Emerging and Zoonotic Diseases. Vector Borne Zoonotic Dis 2006;6:14-23.
16. Green MS, Kaufman Z. Surveillance for Early Detection and Monioring of Infectious Disease Outbreaks Associated with Bioterrorism. Isr Med Assoc J 2002;4:503-6.
17. Henning KJ. Syndromic Surveillance. Smolinski, Mark S., Hamburg, Margaret A., and Lederberg, Joshua. Microbial Threats to Health: Emergence, Detection, and Response. 2003. Washington, D.C., National Academy Press.
18. Johnson HA, Wagner MM, Saladino RA. A New Method for Investigating Non-traditional Biosurveillance Data: Studying Behaviors Prior to Emergency Department Visits. 2005.

19. Kaufmann AF, Meltzer MI, Schmid GP. The Economic Impact of a Bioterrorist Attack: Are Prevention and Postattack Intervention Programs Justifiable? Emerg Infect Dis 1997;3:83-94.
20. Kearney B. Strengthening Safeguards Against Disease Outbreaks. In Focus 5[2]. 2005. Washington, D.C., The National Academy of Sciences.
21. Kelsey H. Improvements in methodologies for tracking infectious disease needed. The Newsbulletin . January 13, 2005. Los Alamos National Laboratory.
22. Lucas JM. Counted Data CUSUM's. Technometrics 1985;27:129-44.
23. Mauer WA, Kaneene JB. Integrated Human-Animal Disease Surveillance. Emerg Infect Dis 2005;11:1490-1.
24. Moodie M et al. Biological Terrorism in the United States: Threat, Preparedness, and Response. November 2000. Chemical and Biological Arms Control Institute.
25. National Research Council. Animal Health at the Crossroads: Preventing, Detecting, and Diagnosing Animal Diseases. July 2005. Washington, D.C., The National Academy of Sciences.
26. Power C. Passive Animal Disease Surveillance in Canada: A Benchmark. Proceedings of a CAHNet Workshop. November 1999. Canadian Food Inspection Agency.
27. Shaffer LE et al. Evaluation of Microbiology Orders from two Veterinary Diagnostic Laboratories as Potential Data Sources for Early Outbreak Detection. Adv Disease Surveil. forthcoming.
28. Shephard R, Aryel RM, Shaffer L. Animal Health. In: Wagner MM, Moore AW, Aryel RM, eds. Handbook of Biosurveillance. New York, NY: Elsevier Inc., 2006:111-27.
29. Sosin DM. Draft Framework for Evaluating Syndromic Surveillance Systems. J Urban Health 2003;80:i8-i13.
30. Tsui F-C et al. Value of ICD-9-Coded Chief Complaints for Detection of Epidemics. Proceedings of the AMIA Annual Symposium. 711-715. 2001.
31. Vourc'h G et al. Detecting Emerging Diseases in Farm Animals through Clinical Observations. Emerg Infect Dis 2006;12:204-10.
32. Wagner MM, Aryel R, Dato V. Availability and Comparative Value of Data Elements Required for an Effective Bioterrorism Detection System. November 28, 2001. Agency for Healthcare Research and Quality.
33. Witt CJ. Electronic Surveillance System for the Early Notification of Community-based Epidemics (ESSENCE). December 11, 2003. Department of Defense Global Emerging Infections System.
34. Woolhouse MEJ . Population biology of emerging and re-emerging pathogens. Trends Microbiol 2002;10:S3-S7.

Chinese Chief Complaint Classification for Syndromic Surveillance

Hsin-Min Lu[1], Chwan-Chuen King[2], Tsung-Shu Wu[2], Fuh-Yuan Shih[3],
Jin-Yi Hsiao[2], Daniel Zeng[1], and Hsinchun Chen[1]

[1] Management Information Systems Department, Eller College of Management,
University of Arizona, Tucson, Arizona 85721, USA
hmlu@email.arizona.edu, {zeng, hchen}@eller.arizona.edu
[2] Graduate Institute of Epidemiology, National Taiwan University, Taipei, Taiwan
cc_king99@hotmail.com, wcsg@msn.com, pingeshity@yahoo.com.tw
[3] Department of Emergency Medicine, National Taiwan University Hospital, No. 7,
Chung-Shan South Road, Taipei 100, Taiwan

Abstract. There is a critical need for the development of chief complaint (CC) classification systems capable of processing non-English CCs as syndromic surveillance is being increasingly practiced around the world. In this paper, we report on an ongoing effort to develop a Chinese CC classification system based on the analysis of Chinese CCs collected from hospitals in Taiwan. We found that Chinese CCs contain important symptom-related information and provide a valid source of information for syndromic surveillance. Our technical approach consists of two key steps: (a) mapping Chinese CCs to English CCs using a mutual information-based mapping method, and (b) reusing existing English CC classification systems to process translated Chinese CCs. We demonstrate the effectiveness of this proposed approach through a preliminary evaluation study using a real-world dataset.

Keywords: multilingual chief complaint classification, Chinese chief complaints, syndromic surveillance, medical ontology, UMLS, mutual information.

1 Introduction

Modern transportation has shortened the time needed for a person to travel from one side of the globe to the other. At the same time, it also shortened the time needed for a disease to spread. A case in point is the severe acute respiratory syndrome (SARS) episode which started in the Guangdong Province, China in November, 2002 and spread to Toronto, Vancouver, Ulaan Bator, Manila, Singapore, Hanoi, and Taiwan by March, 2003. The disease was finally brought under control and the whole episode ended in July, 2003. There were a total of 8096 known cases, and about 35% were outside mainland China.[1]

The SARS experience indicates that an effective plan for infectious disease detection and prevention, in which syndromic surveillance may play an important

---

[1] http://www.who.int/csr/sars/en/

D. Zeng et al. (Eds.): BioSurveillance 2007, LNCS 4506, pp. 11–22, 2007.

role, should be considered on a global scale. However, only a few countries have adopted formal syndromic surveillance systems. The U.S. public health system has significant experience in developing and adopting syndromic surveillance systems. However, leveraging such experience in international contexts proves to be difficult. Multilingual data presents a major barrier, as different languages are used by medical and public health practitioners in different parts of the world. This is particularly true for a major source of data used by many syndromic surveillance systems: emergency department (ED) triage free-text chief complaints (CCs).

ED triage free-text chief CCs are short free-text phrases entered by triage practitioners describing reasons for patients' ED visits. ED CCs are a popular data source because of their timeliness and availability [1-4]. However, medical practitioners in other countries do not always use English when recording patients' CCs. As a result, existing CC classification systems designed for English CCs cannot be directly applied in these countries as an important component of the overall syndromic surveillance strategy.

This paper reports an ongoing study on the role of Chinese CCs as a data source for syndromic surveillance and the related systems work to develop a Chinese CC syndromic classification approach. CCs from EDs in Taiwan were collected and analyzed. Medical practitioners in Taiwan are trained to record CCs in English. However, it is a common practice to record CCs in both Chinese and English phrases. Furthermore, some hospitals record CCs only in Chinese. As an initial step of our research program, we systematically investigated the role and validity of Chinese CCs in the syndromic surveillance context. We then developed an initial approach to classify Chinese CCs based on an automated mechanism to map Chinese CCs to English CCs. This paper summarizes the key findings of and lessons learned from this investigation.

The remainder of this paper is organized as follows. Section 2 provides the background for the CC classification problem and articulates the research opportunities and the objective of our research. Section 3 presents our research findings about using Chinese CCs as a data source. Section 4 reports our initial design for a Chinese CC classification system and the results of a preliminary evaluation study. The last section summarizes our research and discusses future directions.

2   Research Background

This section reviews existing English and non-English CC classification research. English CC classification typically uses one of three major methods: supervised learning, rule-based, and ontology-enhanced. Non-English CC classification research is still in its early stage of development. A popular method is to classify CCs based on keyword matching.

2.1 English Chief Complaint Classification Methods

There are three main approaches for automated CC syndrome classification: supervised learning, rule-based classification, and ontology-enhanced classification. The supervised learning methods require CC records to be labeled with syndromes before being used for model training. Naive Bayesian [5, 6] and Bayesian network [2] models are two examples of the supervised learning methods studied. Implementing the learning algorithms is straightforward; however, collecting training records is usually costly and time-consuming. Another major disadvantage of supervised learning methods is the lack of flexibility and generalizability. Recoding for different syndromic definitions or implementing the CC classification system in an environment which is different from the one where the original labeled training data were collected could be costly and difficult.

Rule-based classification methods use a completely different approach and do not require labeled training data. Such methods typically have two stages. In the first stage, CC records are cleaned up and transformed to an intermediate representation called "symptom groups" by either a symptom grouping table (SGT) lookup or keyword matching. In the second stage, a set of rules is used to map the intermediate symptom groups to final syndromic categories. For instance, the EARS[2] system uses 42 rules for such mappings.

A major advantage of rule-based classification methods is their simplicity. The syndrome classification rules and intermediate SGTs can be constructed using a top-down approach. The "white box" nature of these methods makes system maintenance and fine-tuning easy for system designers and users. In addition, these methods are flexible. Adding new syndromic categories or changing syndromic definitions can be achieved relatively easily by switching the inference rules; the SGTs can typically be shared across hospitals.

A major problem with rule-based classification methods is that they cannot handle symptoms that are not included in the SGTs. For example, a rule-based system may have a SGT containing the symptoms "abdominal pain" and "stomach ache" which belong to the symptom group "abd_pain." This system, however, will not be able to handle "epigastric pain" even though "epigastric pain" is closely related to "abdominal pain."

The BioPortal CC classifier [7] is designed to address this vocabulary problem using an ontology-enhanced approach. The semantic relations in the Unified Medical Language System (UMLS), a medical ontology, are used to increase the performance of a rule-based chief complaint classification system. At the core of this approach is the UMLS-based weighted semantic similarity score (WSSS)-based grouping method that is capable of automatically assigning symptoms previously un-encountered to appropriate symptom groups. An evaluation study shows that this approach can achieve a higher level of sensitivity, F measure, and F2 measure when compared to a rule-based system that has the same symptom grouping table and syndrome rules.

2.2 Non-english Chief Complaint Classification Methods

Little attention has focused on non-English CC classifications. One straightforward extension is adding non-English keywords into existing English CC classification systems. For instance, this approach has been applied to process Spanish CCs in EARS [8]. However, in languages without clear word boundaries such as Chinese or Japanese, this method is not applicable.

---

[2] http://www.bt.cdc.gov/surveillance/ears/

It is possible to use the ICD-9 codes instead of free-text CCs to classify ED records. For example, Wu et al. used ICD-9 codes attached to ED records to classify Chinese CCs into eight syndromic categories [9]. However, as ICD-9 codes are primarily used for billing purposes, they are not always informative for syndromic surveillance [10, 11]. As such, free-text CCs remain one of the most important data sources for syndromic surveillance [12].

3   The Role of Chinese Chief Complaints

The long-term objective of our research is to fill in the technical gaps existing in the current multilingual CC classification research and develop practical automatic syndromic classification approaches that can handle both English and non-English CCs. As the first step towards this objective, we have conducted a case study to investigate the prevalence and usefulness of Chinese CCs in the syndromic surveillance context based on a dataset collected from a number of hospitals in Taiwan.

Our working definition of Chinese CCs is any CC records containing Chinese characters. Specialized punctuation marks, which belong to standard-compliant computerized Chinese character sets, are also considered as Chinese characters. In order to validate Chinese CCs as an input to syndromic surveillance systems, we have developed a computer program to calculate the prevalence of Chinese CCs and select random samples from our dataset for further analysis to better understand their role. This section reports on the data collection effort, followed by a discussion of our experimental design and findings.

3.1   Chinese Chief Complaint Dataset

The Chinese CC dataset used in our study consists of about 2.4 million chief complaint records from 189 hospitals in Taiwan. About 98% of these records have admission times from Jan. 1, 2004 to Jan. 26, 2005. Excluding records containing null CCs, there are 939,024 CC records which have been investigated in this study.

3.2   Experimental Design

Manual evaluation of the nearly one million records in our Chinese CC dataset would be impractical. Our experimental investigation followed a two-step design. In the first step, a computer program was used to distinguish whether a CC record contains Chinese characters. The prevalence of Chinese CCs can then be calculated from the output of the program.

Since the focus of this study is to understand the usefulness of Chinese CCs for syndromic surveillance, in the second step we focus on the hospitals that have more than 10% of their CC records containing Chinese characters. For each hospital meeting this threshold, a random sample of 30 Chinese CC records was drawn for manual review. In total, 20 hospitals met this condition and were reviewed. The 600 records from these 20 hospitals were then merged in a random order.

A coder read through all 600 records and classified the Chinese components of the records into four major categories: symptom-related, name entity, Chinese punctuation, and other. Two examples for the CC records belonging to the first "symptom-related" category are " 今早開始腹痛; 剛吃藥後始雙眼腫, 現呼吸不適, 心悸 (verbatim translation: abdominal pain from this morning; eyes swollen after taking medication, shortness of breath, palpitations)" and "昨天開始腹瀉 (verbatim translation: diarrhea started yesterday)." From time to time, triage nurses may find it hard to translate names of places, people, restaurants, among others, and as a result, keep them in Chinese while still describing symptoms in English. For example, in the CC record "Diarrhea SINCE THIS MORNING. Group poisoning. Having dinner at 喜滿客 restaurant," the restaurant name was kept in Chinese while everything else was in English. This set of CC records is classified as "name entity." The third category, Chinese punctuation, consists of CCs with English phrases and Chinese punctuation marks. For example, the record "FEVER SINCE YESTERDAY, COUGH FOR 3-4 DAYS -THROAT INJECTED, LUNG:BS CLEAR" is actually an English CC. But the nurse used the comma available from the Chinese character set " , " instead of the comma symbol "," commonly used in English sentences. This may be due to the keyboard input method used by some hospitals. Finally, CCs that do not belong to any of these three categories were coded as other.

3.3 Experimental Results

Table 1 summarizes the prevalence of Chinese CCs. The overall prevalence of Chinese CCs in the entire Taiwan CC dataset is about 25%. Among the types of hospitals covered by this dataset, medical centers have the highest prevalence rate of 52%, followed by local hospitals (19%), and regional hospitals (16%). The hospital with the highest prevalence at the medical center level is the Mackey Memorial Hospital (馬偕紀念醫院), which has 100% of its CC records in Chinese. The hospital with the second highest prevalence is the National Taiwan University Medical Center (台大醫院) with a prevalence of 18%.

Table 1. Chinese chief complaint prevalence

	# Records	# Hospitals	% Chinese CCs
Medical Center	222,893	17	52%
Regional Hospital	484,123	57	16%
Local Hospital	232,008	89	19%
Total	939,024	163	25%

Table 2 summarizes the results of the analysis performed in the second step of our study. There are 20 hospitals with Chinese CC prevalence higher than 10%. The second row of Table 2 reports the percentages of each of the four target categories, averaged across all 20 hospitals. The third row reports similar percentages averaged across all hospitals but weighted by the size of each hospital, measured by the total number of CCs from each hospital. These results demonstrate that more than half of the time, the Chinese CC records contain information related to symptoms.

Table 2. The categories of Chinese expressions in Chinese chief complaints

Category	Symptom-Related	Name Entity	Chinese Punctuation	Other
Simple Average[*]	40.79%	13.97%	20.32%	24.92%
Weighted Average[**]	53.80%	7.36%	14.63%	14.63%

[*] Equally weighed for all hospitals.
[**] Weighed by the number of Chinese CC records at each hospital.

4 A Chinese Chief Complaint Classification Approach

The study reported above indicates the importance of Chinese CCs as a valid data source for syndromic surveillance. This section reports our preliminary work on designing and evaluating a CC classification system that can process both Chinese and English CCs.

It is possible to develop a Chinese CC classification approach from scratch. However, since there are many effective CC classification methods already developed for English CCs, it is worthwhile exploring the possibility of reusing these methods and extending them to handle Chinese CC records. A prerequisite of this attempt is that the Chinese expressions in Chinese CCs must be translated into English. There are several multilingual text processing techniques that can be applied in this context. In this section, we present our effort to apply one such technique based on mutual information to translate Chinese CCs into English and extend the BioPortal CC classification system, which was developed in our prior research for English CCs, to handle Chinese CCs.

4.1 Chinese Chief Complaint Preprocessing

Translating Chinese CCs to English is different from general translation tasks. The Chinese expressions in CCs are in most cases words or phrases as opposed to complete sentences. Moreover, not every word or phrase is informative for syndromic surveillance purposes. As a result, the purpose of Chinese CC preprocessing is not verbatim translation of Chinese expressions contained in CC records. Instead, only information that is useful and relevant to syndromic surveillance is *extracted* from the original Chinese CCs.

To meet this requirement, instead of using a general Chinese-English dictionary or a machine translation program to extract symptom-related information, we develop a new approach centered around a Chinese-English symptom mapping table based on mutual information. In this approach, Chinese symptom-related terms are collected and their mappings to English expressions are created. Words or phrases not listed in the mapping table are ignored in the Chinese CC preprocessing step. Below we present in detail how this mutual information-based mapping (MIM) table is constructed.

First, a symptom phrase list is extracted from the sample CC records (a testing dataset) based on the co-occurrence pattern as follows. We define the mutual information of a phrase as

$$MI_c = \frac{P(c)}{P(a)+P(b)-P(c)} \tag{1}$$

where c is the pattern of interest (e.g., 人工智慧; artificial intelligence); a and b are left and right subpatterns of c, i.e. a= 人工智 (a partial word without meaning) and b= 工智慧 (a partial word without meaning). Based on this measure, MI_c will be substantially higher than other random patterns if c is by itself a phrase and its subpattern a and b appears in text only because of c. For instance, c=″人工智慧″ may appear in text 9 times. Its subpattern a=″人工智″ and b=″工智慧″ appear in the text only because it is the subpattern of c. In this case, we have MI=9/(9+9-9)=1. Thus we can determine from the mututal information that c should be considered as a phrase.

A phrase extraction tool based on mutual information [13] read all CCs in the testing dataset and outputted a phrase list. Four hundred and fifteen symptom-related words and phrases were extracted manually from the original phrase list. This symptom phrase list, together with a general Chinese phrase list, was used to perform Chinese word segmentation. The segmentation results on the testing dataset were then reviewed and additional phrases were included in the symptom phrase list. The final symptom phrase list consists of 470 Chinese phrases.

After identifying these Chinese phrases, three physicians helped us translate them into English. We provided the physicians with a spreadsheet listing the Chinese symptom phrases together with example CCs which contain these phrases. We then reviewed the translations from these physicians to make sure that translations are consistent and meaningful. The final MIM table consists of 380 Chinese-English mappings.

4.2 A System Design for Chinese Chief Complaint Processing

Fig. 1 depicts the design of our Chinese CC classification approach based on the mappings discussed above. In this approach, Chinese CC processing follows four major stages. Stages 0.1 to 0.3 separate Chinese and English expressions in CCs, perform word/phrase segmentation for Chinese expressions, and map symptom-related phrases to English. At the end of these stages, CC records are in English. In the follow-up stages (1-3) the BioPortal English CC classifier is invoked [7] and produces the final syndrome classification results. Below we discuss each of these steps in detail.

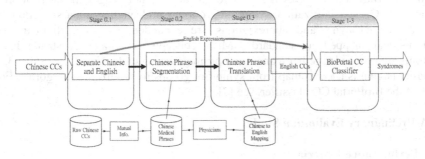

Fig. 1. Chinese chief complaint classification

4.2.1   Stage 0.1: Separating Chinese and English Expressions

Stage 0.1 separates Chinese from English expressions. Since the BioPortal classifier can process English CCs, in this stage any existing English terms are kept. As part of processing the Chinese expressions, their position relative to English terms is also marked and recorded. In a later stage, the mapped English expressions will replace their original Chinese counterparts. For example, the chief complaint record "Dyspnea,SOB早上開始坐骨神經痛 解尿困難" is first divided into two parts: "Dyspnea,SOB," which will skip the Chinese term preprocessing steps, and "早上開始坐骨神經痛 解尿困難," which will be sent to Stage 0.2 for word/phrase segmentation.

4.2.2   Stage 0.2: Chinese Expression Segmentation

In this stage, Chinese expressions in the Chinese CCs are segmented using the Chinese terms available from the MIM table. The longest possible phrases in the phrase list are used for segmentation purposes. For example, the expression "與人打架, 用鍋鏟打到頭部, 流鼻血 (verbatim translation: fight with man, hit in head by a Chinese spatula, epistaxis)" will be segmented as " 與-人-打架 , 用-鍋鏟-打到頭-部 , 流鼻血 (verbatim translation: with-man-fight, use-Chinese spatula-head injury-part, epistaxis)."

4.2.3   Stage 0.3: Chinese Phrase Translation

The segmented phrases/words that are the output of the previous step are used as the basic elements for Chinese-English symptom mapping. The MIM table is consulted to obtain the final translation results. For example, the segmented Chinese expression " 與-人-打架 , 用-鍋鏟-打到頭-部 , 流鼻血" is mapped to the following English expressions: "N/A-N/A-fighting , N/A-N/A-head injury-N/A , epistaxis." "N/A" indicates a term unavailable in the mapping table. The final translated result thus is "fighting, head injury, epistaxis."

4.2.4   Stages 1-3: Connecting to BioPortal Chief Complaint Classification System

After substituting the translated English expressions for the processed Chinese counterparts in the CCs, we are ready to obtain the final classification results by invoking the BioPortal CC classifier. There are three major Stages in the BioPortal CC classifier: CC standardization, symptom grouping, and syndrome classification. In Stage 1, the acronyms and abbreviations are expanded. CCs are divided into symptoms and mapped to standard UMLS concepts. In Stage 2, standardized symptoms are grouped together using a symptom grouping table. Finally, in Stage 3, a rule-based engine is used to map from symptom groups to syndromic categories. For details of the BioPortal CC classifier, see [7].

4.3   A Preliminary Evaluation Study

4.3.1   Performance Criteria

Sensitivity (recall), specificity (negative recall), and positive predictive value (PPV or precision) have all been used extensively in previous research [2, 3, 5]. In addition to the three criteria mentioned above, we also consider the F measure and F2 measure [14, 15]. The F and F2 measures, commonly used in the information retrieval literature, combine PPV and sensitivity to provide a single integrated measure to evaluate the overall performance of a given approach.

Formally, specificity, sensitivity, and PPV are defined as

$$Specificity = \frac{TN}{FP + TN} \tag{2}$$

$$Sensitivity = \frac{TP}{TP + FN} \tag{3}$$

$$PPV = \frac{TP}{TP + FP} \tag{4}$$

where TP is the number of true positive cases, FN the number of false negative cases, TN the number of true negative cases, and FP the number of false positive cases.

The F measure is a weighted harmonic mean of PPV and sensitivity. In the context of syndromic surveillance, sensitivity is more important than precision and specificity [2]. The F2 measure gives recall twice as much weight as precision and thus can reflect the emphasis on recall. These two measures are defined as follows:

$$F = \frac{2PPV \times Sensitivity}{PPV + Sensitivity} \tag{5}$$

$$F2 = \frac{(1 + 2)PPV \times Sensitivity}{2PPV + Sensitivity} \tag{6}$$

4.3.2 Syndromic Definitions and Reference Standard

In our evaluation study, we use 8 syndrome categories. These categories were chosen at a syndromic surveillance meeting by five physicians in Taiwan. They are: constitutional, gastrointestinal, rash, respiratory, upper respiratory, lower respiratory, fever, and other. "Other" is a miscellaneous category for CCs that do not fit into any of the other syndromes. CCs assigned to "other" syndrome were not considered in the subsequent evaluation study. One chief complaint could be assigned to more than one syndrome. For example, if upper respiratory or lower respiratory is assigned, the respiratory syndrome automatically applies as well.

To the best of our knowledge, there are no publicly available labeled Chinese CCs with compatible syndrome definitions. Therefore the reference standard for the system evaluation had to be constructed for this study. The testing dataset is a random sample of 300 CC records from the Mackey Memorial Hospital in Taiwan. CCs with exactly the same contents are removed before random sampling. The syndrome classifications for these records were provided by a practitioner.

We observe that one CC is on average assigned to 1.39 syndromes. The gastrointestinal syndrome has the highest prevalence of 30.00%. About 23.67% of

CCs contain a fever symptom. Constitutional and respiratory syndromes have prevalence of about 12.67% and 17.33%, respectively.

4.3.3   Experimental Results

The performance of our proposed approach, called mutual information-based mapping (MIM), can be found in Table 3. The second column of Table 3 lists the valid cases in each syndromic category. The third through fifth columns list the PPV, sensitivity, specificity, F, and F2 measures. We can see that most syndromes have F and F2 measures higher than 0.9, indicating promising performance. The two exceptions are the constitional syndrome and the rash syndrome, which have F measures of 0.3511 and 0.6000, respectively. The PPV and sensitivity shows similar patterns. Most syndromes have PPV higher than 0.9, indicating that more than 90% of records marked by the Chinese CC classification system are correct. Sensitivity is also reasonably good, which means that the system can catch most positive cases in the reference standard. Specificity of most syndromes is close to one because 1) most syndromes have low prevalence and 2) the system in general can successfully mark negative cases. Overall, the Chinese CC classification system shows promising performance.

Table 3. Performance of the MIM approach

Syndrome	TP+ FN	PPV	Sensitivity	Specificity	F	F2
CONST	38	0.2473	0.6053	0.7328	0.3511	0.4083
GI	90	0.9775	0.9667	0.9905	0.9721	0.9703
RASH	6	0.7500	0.5000	0.9966	0.6000	0.5625
RESP	52	0.9259	0.9615	0.9839	0.9434	0.9494
URESP	20	0.9091	1.0000	0.9929	0.9524	0.9677
LRESP	42	0.9524	0.9524	0.9922	0.9524	0.9524
FEVER	71	1.0000	0.9859	1.0000	0.9929	0.9906

Comparing the performance among syndromic categories, fever symptom has the best performance with PPV, sensitivity, and specificity of nearly one. Analyzing the original free-text CCs, we found that CCs describing fever symptoms usually contain specific keywords (e.g. "發燒," "發高燒"). Respiratory syndrome, which has more related phrases, also performs well. On the other hand, constitutional syndrome has a broader definition and variety of expressions. As a result, it is possible that constitutional syndrome has the lowest performance because MIM did not provide a good mapping. It is also possible that the definition of constitutional syndrome is not clear and the reference standard is not consistent. Expanding the size of reference standard and collecting syndrome assignments from multiple experts should help to isolate the problem.

5 Concluding Remarks

To summarize, Chinese CCs provide a valid data source for syndromic surveillance. To take advantage of Chinese CC records, we have developed a Chinese CC classification approach that can handle both Chinese and English. Preliminary experimental results show that our approach performs well for most syndromic categories. Our current work is focused on deriving a larger reference standard for performance evaluation and comparing our approach with alternative Chinese-English translation approaches.

Acknowledgement

This work was supported in part by the U.S. National Science Foundation through Grant #IIS-0428241. It draws on earlier work supported by the Arizona Department of Health Services. Chwan-Chuen King acknowledges support from the Taiwan Department of Health (DOH95-DC-1021). Daniel Zeng is an affiliated professor at the Institute of Automation, the Chinese Academy of Sciences, and wishes to acknowledge support from a research grant (60573078) from the National Natural Science Foundation of China, an international collaboration grant (2F05N01) from the Chinese Academy of Sciences, and a National Basic Research Program of China (973) grant (2006CB705500) from the Ministry of Science and Technology.

References

1. Chapman, W.W., Dowling, J.N., Wagner, M.M.: Generating a Reliable Reference Standard Set for Syndromic Case Classification. Journal of the American Medical Informatics Association 12 (2005) 618-629
2. Chapman, W.W., Christensen, L.M., Wagner, M.M., Haug, P.J., Ivanov, O., Dowling, J.N., Olszewski, R.T.: Classifying Free-Text Triage Chief Complaints into Syndromic Categories with Natural Language Processing. Artificial Intelligence in Medicine 33 (2005) 31-40
3. Ivanov, O., Wagner, M.M., Chapman, W.W., Olszewski, R.T.: Accuracy of Three Classifiers of Acute Gastrointestinal Syndrome for Syndromic Surveillance. AMIA Symposium (2002) 345-349
4. Espino, J.U., Wagner, M.M.: The Accuracy of Icd-9 Coded Chief Complaints for Detection of Acute Respiratory Illness. Proceedings of the AMIA Annual Symposium (2001) 164-168
5. Olszewski, R.T.: Bayesian Classification of Triage Diagnoses for the Early Detection of Epidemics. FLAIRS Conference Menlo Park, California (2003) 412-416
6. Espino, J.U., Dowling, J., Levander, J., Sutovsky, P., Wagner, M.M., Copper, G.F.: Syco: A Probabilistic Machine Learning Method for Classifying Chief Complaints into Symptom and Syndrome Categories. Syndromic Surveillance Conference, Baltimore, Maryland (2006)
7. Lu, H.-M., Zeng, D., Chen, H.: Ontology-Enhanced Automatic Chief Complaint Classification for Syndromic Surveillance. MIS Department, University of Arizona (2006)

8. Hutwagner, L., Thompson, W., Seeman, G.M., Treadwell, T.: The Bioterrorism Preparedness and Response Early Aberration Reporting System (Ears). Journal of Urban Health 80 (2003) i89-i96
9. Wu, T.-S.: Establishing Emergency Department-Based Infectious Disease Syndromic Surveillance System in Taiwan–Aberration Detection Methods, Epidemiological Characteristics, System Evaluation and Recommendations. Graduate Institute of Epidemiology, Vol. Master Thesis. National Taiwan University, Taipei (2005)
10. Fisher, F.S., Whaley, F.S., Krushat, W.M., Malenka, D.J., Fleming, C., Baron, J.A., Hsia, D.C.: The Accuracy of Medicare's Hospital Claims Data: Progress Has Been Made, but Problems Remain. American Journal of Public Health 82 (1999) 243-248
11. Day, F.C., Schriger, D.L., La, M.: Automated Linking of Free-Text Complaints to Reason-for-Visit Categories and International Classification of Diseases Diagnoses in Emergency Department Patient Record Databases. Annals of Emergency Medicine 43 (2004) 401-409
12. Yan, P., Chen, H., Zeng, D.: Syndromic Surveillance Systems: Public Health and Biodefence. Annual Review of Information Science and Technology forthcoming (2006)
13. Ong, T.-H., Chen, H.: Updateable Pat-Tree Approach to Chinese Key Phrase Extraction Using Mutual Information: A Linguistic Foundation for Knowledge Management. Proceedings of the Second Asian Digital Library Conference, Taipei, Taiwan (1999)
14. van Rijsbergen, C.J.: Information Retrieval, Butterworths, London (1979)
15. Pakhomov, S.V.S., Buntrock, J.D., Chute, C.G.: Automating the Assignment of Diagnosis Codes to Patient Encounters Using Example-Based and Machine Learning Techniques. Journal of the American Medical Informatics Assocation 13 (2006) 516-525

Incorporating Geographical Contacts into Social Network Analysis for Contact Tracing in Epidemiology: A Study on Taiwan SARS Data

Yi-Da Chen[1], Chunju Tseng[1], Chwan-Chuen King[2],
Tsung-Shu Joseph Wu[2], and Hsinchun Chen[1]

[1] Department of Management Information Systems, The University of Arizona
Tucson, AZ 85721, USA
{ydchenb, chunjue, hchen}@eller.arizona.edu
[2] Graduate Institute of Epidemiology, National Taiwan University
Taipei, Taiwan

Abstract. In epidemiology, contact tracing is a process to control the spread of an infectious disease and identify individuals who were previously exposed to patients with the disease. After the emergence of AIDS, Social Network Analysis (SNA) was demonstrated to be a good supplementary tool for contact tracing. Traditionally, social networks for disease investigations are constructed only with personal contacts. However, for diseases which transmit not only through personal contacts, incorporating geographical contacts into SNA has been demonstrated to reveal potential contacts among patients. In this research, we use Taiwan SARS data to investigate the differences in connectivity between personal and geographical contacts in the construction of social networks for these diseases. According to our results, geographical contacts, which increase the average degree of nodes from 0 to 108.62 and decrease the number of components from 961 to 82, provide much higher connectivity than personal contacts. Therefore, including geographical contacts is important to understand the underlying context of the transmission of these diseases. We further explore the differences in network topology between one-mode networks with only patients and multi-mode networks with patients and geographical locations for disease investigation. We find that including geographical locations as nodes in a social network provides a good way to see the role that those locations play in the disease transmission and reveal potential bridges among those geographical locations and households.

Keywords: Social Network Analysis, Contact Tracing, Epidemiology, Personal Contacts, Geographical Contacts, SARS.

1 Introduction

In epidemiology, contact tracing is a process to control the spread of an infectious disease and to identify individuals who were previously exposed to patients with the

D. Zeng et al. (Eds.): BioSurveillance 2007, LNCS 4506, pp. 23–36, 2007.

disease. Through contact tracing, healthcare workers can trace the possible source of infection for those patients, monitor individuals who may develop the disease, and prevent further spread of the disease. Traditional contact tracing is based on the notion that disease spread is either serial or parallel, but linear in either case [13]. According to this notion, the transmission of infectious diseases is like a branching tree with a single source and no interconnection among leaf nodes. This notion, however, ignores the fact that people are interconnected in a complex social context and that such interconnection has implications for the spread of infectious diseases.

Although the idea of social network analysis in contact tracing can be traced back to the 1930s [1], it got major attention after the emergence of AIDS. In 1985, Klovdahl [6] used AIDS as an example to illustrate the usefulness of Social Network Analysis (SNA) in studying the transmission of an infectious disease. Since then, SNA has been successfully applied to AIDS [8][14], gonorrhea [4], syphilis [15], tuberculosis (TB) [7][9], and Severe Acute Respiratory Syndrome (SARS) [10].

Through those studies, SNA has been demonstrated to be a good supplementary tool for contact tracing. Compared to traditional contact tracing, SNA provides a perspective to conceptualize the disease transmission within a group of people, identify the people who act as bridges between subgroups, and reveal some previously unrecognized patterns of transmission. In a study of a TB outbreak, McElroy et al. [9] found that the use of crack cocaine was an important factor in the outbreak which had not been recognized by the local TB control program.

Since personal contacts are the most identifiable paths for disease transmission, the majority of SNA studies in epidemiology use personal contacts to model the spread of diseases. For sexually transmitted diseases (STDs), which require intimate contact for transmission, personal contacts are adequate to form a well-connected network for investigation. However, for diseases which can also transmit through the air or through contaminated inanimate objects, such as TB and SARS, personal contacts alone are not enough to explain the underlying context of disease transmission. In a study of a TB outbreak in Houston, Texas, Klovdahl et al. [7] included places as a type of actors in SNA and found that geographical contacts were critical for understanding the outbreak. In another study of a TB outbreak, McElroy et al. [9] also included attendance at exotic dance clubs in their network analysis and discovered some potential contacts among patients.

Incorporating geographical contacts into SNA for contact tracing may raise some doubts about the connection of two people via their geographical contacts. Since geographical locations are places of social aggregation, many people are easily connected together via their geographical contacts without any actual contact. However, from these two studies, we can see that incorporating geographical contacts into SNA provides us a good way to find potential connections among patients and to see the role that those geographical locations play in disease outbreaks. Therefore, in this research, we aim to further investigate the necessity of incorporating geographical contacts into SNA for contact tracing and explore the strengths of multi-mode networks with patients and geographical locations for disease investigation.

The remainder of the paper is organized as follows. In Section 2, we review the studies of SNA in epidemiology. In Sections 3 and 4, we discuss our research questions and present our research design. In Section 5, we report our analysis results. Section 6 concludes this paper with implications and future directions.

2 Literature Review

In epidemiology, the modeling of disease transmission traditionally focuses on biological factors, such as the period of infectiousness and duration of incubation, and treats social factors as random or homogeneous mixing [11]. After the emergence of AIDS, SNA was shown to provide better understanding of the spread of some infectious diseases which transmit through intimate personal contact [6].

SNA has two major steps: network construction and analysis.

2.1 Network Construction

During network construction, individuals in a group are represented as nodes in a network and connected through certain types of contact. Table 1 shows the taxonomy of the network construction. When SNA is applied to investigate disease transmission, the types of contact used to construct social networks should resemble the paths of the transmission. For STDs, a sexual contact network is the major focus for investigation. In 1985, Klovdahl [6] used the AIDS data from Centers for Disease Control and Prevention (CDC) to illustrate how sexual contact could link 40 patients together. In 1994, Klovdahl et al. [8] further applied SNA to investigate an AIDS outbreak in Colorado Springs and connected over 600 individuals in a social network with multiple types of contact, including sexual contact, drug use, needle sharing, and social contact.

Table 1. Taxonomy of Network Construction

Network Construction		
Disease	**Type of Contacts**	**Example**
Sexually Transmitted Disease (STD)	Sexual Contact	AIDS (Klovdahl, 1985) AIDS (Klovdahl et al., 1994)
	Drug Use	AIDS (Klovdahl et al., 1994)
	Needle Sharing	AIDS (Klovdahl et al., 1994)
	Social Contact	AIDS (Klovdahl et al., 1994)
Tuberculosis (TB)	Personal Contact	(Klovdahl et al., 2001) (McElroy et al., 2003)
	Geographical Contact	(Klovdahl et al., 2001) (McElroy et al., 2003)
Severe Acute Respiratory Syndrome (SARS)	Source of Infection	(CDC, 2003) (Shen et al. 2004)

In a study of a TB outbreak in Houston, Texas, Klovdahl et al. [7] included geographical contacts in their network and found that they were critical for understanding that outbreak. Later McElroy et al. [9] also included geographical contacts and had similar findings in their investigation.

During the SARS outbreak in 2003, there were several super-spreaders, patients who directly infected more than 10 other people, reported in Singapore [2] and Beijing [16]. To illustrate SARS transmission through super-spreaders, social networks were used and constructed via the reported source of patients' infection.

2.2 Network Analysis

There are three levels of network analysis: network visualization, measurement, and simulation.

2.2.1 Network Visualization
In the visualization level, SNA is applied to visualize the disease transmission within a particular group. Through network visualization, we can identify subgroups within the group and the people who act as bridges to transmit the investigated disease from one subgroup to another. Combining some background information, we can further characterize each group and explain the underlying context of the transmission.

In their investigation of a syphilis outbreak, Rothenberg et al. [15] reported that network visualization uncovered several people who acted as bridges between subgroups. They also included ethnographical data in their analysis and indicated that a group of young white girls was at the center of the outbreak and had interactions with several groups of boys, including a group of African-American boys.

Network visualization for disease investigation suffers from some limitations. First, if the inputted network data is incomplete, the constructed network may have some missing nodes or fragments which could mislead our interpretation of the investigated scenario [4]. Second, the constructed network alone cannot provide a complete picture of the disease transmission [13]. We need some complementary data, such as ethnographic data, to explain the context of environment.

2.2.2 Network Measurement
In the measurement level, we measure structural properties of the constructed networks and make inferences about disease transmission from those properties. In epidemiology, the measurement focuses on the centrality of networks, which provides information on the importance of individuals in a network and reveals potential bridges between subgroups. By understanding the network structure, we can identify the group of people who are central to the disease transmission and design effective containment strategies to break the chains of transmission [6].

Most of the studies of SNA construct and measure only one social network. However, some studies construct several networks at different points of time in order to investigate the changes of network structure through time. For example, in a study of AIDS transmission, Rothenberg et al. [14] constructed three social networks with the same patients at different points of time and assessed network stability, behavioral change, and structural change among these three networks.

2.2.3 Network Simulation
In network simulation, mathematical models are used to form a contact network in which several parameters of the network may influence the transmission of the investigated disease. Network simulation is particularly useful when we have only a little knowledge about the investigated disease. For example, Meyers et al. [10] used three different networks, urban, power law, and Poisson, to study how contact patterns influence SARS transmission and found that in the same network settings, different contact patterns yield totally different epidemiological outcomes.

2.3 Incorporating Geographical Contacts into SNA

The studies of SNA in epidemiology primarily use only personal contacts to construct social networks. This may stem from the fact that personal contacts are the most common and identifiable paths for disease transmission. However, for some diseases which transmit through multiple vectors, personal contacts alone may not be sufficient to explain the underlying context of the transmission. In an investigation of a TB outbreak in Houston, Texas, Klovdahl et al. [7] reported that in 37 investigated patients there were only 12 personal contacts found to explain that outbreak. Therefore, they incorporated geographical contacts into SNA by including geographical locations as nodes in their network and found that geographical contacts were critical for understanding the outbreak. In a study of another TB outbreak, McElroy et al. [9] included attendance at exotic dance clubs in their investigation and found some potential contacts among patients.

Incorporating geographical contacts into SNA for contact tracing may raise some doubts about the connection of patients via their geographical contacts. First, geographical locations are places of social aggregation and it is easy to find many potential contacts among patients through their geographical contacts. Second, it is also questionable to connect two patients together only because they have been to the same place. They may have been to the place in different months. However, as noted by Klovdahl et al. [7], "the appropriate relative person/place emphasis would depend on characteristics of an outbreak and the populations at risk." For the diseases which transmit not only through personal contacts and may have hospital or community outbreaks, incorporating geographical contacts into investigations would pinpoint the role that those geographical locations play in disease transmission.

3 Research Gaps and Questions

Since geographical locations are places of social aggregation, including geographical contacts in SNA helps to find potential connections among patients but also brings some noise into network analysis. Previous studies have not systematically discussed the necessity of incorporating geographical contacts and investigated the differences in connectivity between personal and geographical contacts in the construction of social networks or, more precisely, contact networks for disease investigation.

Previous studies also have not explored the differences in network topology between one-mode networks with only patients and multi-mode networks with patients and geographical locations.

In this research, we aim to answer the following questions:

1. What are the differences in connectivity between personal and geographical contacts in the construction of contact networks for the diseases which transmit not only through personal contacts?
2. What are the differences in network topology between one-mode networks with only patients and multi-mode networks with patients and geographical locations for disease investigation?

4 Research Design

In this section, we first introduce our research test bed which contains Taiwan SARS data and summarize the important events of SARS transmission in 2003. Then we present our research design. All data reported are scrubbed to protect privacy.

4.1 Research Test Bed

SARS first emerged in Guangdong Province, mainland China, in November 2002 [12]. By late February 2003, it spread to more than 20 countries, including Hong Kong, Singapore, Canada, and Taiwan. SARS was transmitted mainly through close personal contacts with infectious droplets. However, there was evidence showing that SARS could also transmit through the air [17] and contaminated inanimate objects [3]. SARS patients were primarily infected in healthcare and hospital settings [12].

The first SARS patient in Taiwan, a businessman working in Guangdong Province, was reported on 8 March 2003 [5]. On 24 April, Taipei Heping Hospital, a municipal hospital, was reported to have the first SARS outbreak in Taiwan. In the beginning of May, several hospitals in Taiwan were reported to have SARS outbreaks.

Taiwan SARS data was collected by the Graduate Institute of Epidemiology at National Taiwan University. It contains the contact tracing data of all SARS patients in Taiwan. In the test dataset, there are 961 patients, including 638 suspected and 323 confirmed SARS patients. The contact tracing data has two main categories, personal and geographical contacts, and nine types of contact. Table 2 summarizes the numbers of records and patients involved in each type of contact.

Table 2. Summary of Research Test Bed

Main Category	Type of Contacts	Record	Suspected Patients	Confirmed Patients
Personal Contacts	Family Member	177	48	63
	Roommate	18	11	15
	Colleague/Classmate	40	26	23
	Close Contact Before the Onset of Symptoms	11	10	12
Geographical Contacts	Foreign Country Travel History	162	100	27
	Hospital Visit	215	110	79
	High Risk Area Visit	38	30	7
	Hospital Admission History	622	401	153
	Workplace (Healthcare Workers)	142	22	120
Total		1425	638	323

4.2 Research Design

Figure 1 shows our research design. We first apply personal and geographical contacts to construct contact networks. Then we perform connectivity analysis and evaluate network topology on the constructed networks.

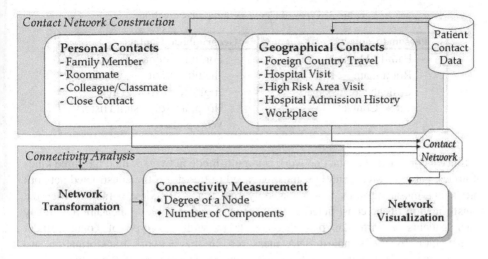

Fig. 1. Research Design

4.2.1 Contact Network Construction

For personal contacts, we represent each SARS patient as a node in a contact network. If two patients have any personal contacts, such as family member, we connect these two patient nodes with a line. For geographical contacts, we introduce areas, such as countries or cities, and hospitals as two additional kinds of node in the contact network. For each geographical contact, we connect the patient node of that contact to the node of the geographical location with a line. Figure 2 demonstrates the contact network construction.

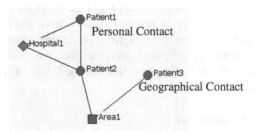

Fig. 2. Example of Contact Network Construction

4.2.2 Connectivity Analysis

Connectivity analysis has two steps: network transformation and connectivity measurement.

4.2.2.1 Network Transformation. In this research, we want to know how many other patients a particular patient is connected with after we apply a certain type of contact to construct a network. We first construct a contact network with only the type of contact we want to investigate. Table 3 lists the types of contact in our investigation.

Table 3. Types of Contacts in the Investigation

Personal Contacts	Geographical Contacts
− Family Member	− Foreign Country Travel
− Roommate	− Hospital Visit
− Colleague/Classmate	− High-Risk Area Visit
− Close Contact	− Hospital Admission History
	− Workplace

Since a constructed contact network is a multi-mode network which has three kinds of nodes, patients, areas, and hospitals, we need to transform the constructed network into a one-mode network which has only one kind of nodes, patients. After the transformation of the constructed contact network, we are able to measure how many other patients a particular patient connects to with the type of contact under investigation by the standard SNA measures. In the network transformation, we connect two patient nodes together in the one-mode network if these two patient nodes are connected to the same geographical node in the contact network. For example, in Figure 2, Patient 2 and 3 are connected to Area 1 in that contact network. In the network transformation, we connect Patients 2 and 3 together in the one-mode network of that contact network since they all have been to Area 1.

4.2.2.2 Connectivity Measurement. In this research, connectivity is defined as the degree to which a type of contact can connect individual patients to form a network. We use two measures commonly used in SNA, the degree of a node and number of components, to measure the connectivity. The degree of a node in a network is the number of lines that are incident with the node. The number of components is the number of maximum connected sub-networks in a network. Logically, if a type of contact has high connectivity in network construction, the degree of a patient node should significantly increase from zero and, at the same time, the number of components should also significantly decrease from the total number of nodes. After each constructed contact network is transformed into a one-mode network, we perform the measurement on it.

4.2.3 Network Visualization

In this research, we also want to know whether including geographical locations as nodes in contact networks provides additional insights in disease investigation. We first evaluate the differences in network topology between a contact network and its transformed one-mode network. Then we analyze the potential insights that a contact network with geographical nodes provides.

5 Research Results

In this section, we present our research results.

5.1 Connectivity Analysis

In the connectivity analysis, we first evaluate the differences in connectivity between two main categories, personal and geographical contacts in the construction of SARS contact networks. Table 4 shows the results of connectivity analysis for personal and geographical contacts. From Table 4, we can see that geographical contacts provide much higher connectivity than personal contacts do in the SARS study. If we only apply personal contacts to construct a contact network, we increase the average degree of nodes from 0 to 0.31 and decrease the number of components from 961, the total number of patient nodes, to 847. In other words, the constructed contact network with only personal contacts is too sparse to get a comprehensive understanding of SARS transmission. In contrast to personal contacts, geographical contacts increase the average degree of nodes from 0 to 108.62 and decrease the number of components from 961 to 82. This means that the majority of SARS patients have no personal contacts with each other and that the connections among them are mainly through attendance at some hospitals or high risk areas. In Table 4, the maximum degree of nodes in geographical contacts is 474. The node with the maximum degree represents a news reporter. He went to two hospitals during the outbreak investigation and then was admitted to another hospital after his onset of symptoms. Therefore, through his geographical contacts he is connected to other patients of these three hospitals.

Table 4. Result of Connectivity Analysis for Personal and Geographical Contacts

	Average Degree	Maximum Degree	Number of Components
Personal Contacts	0.31	4	847
Geographical Contacts	108.62	474	82
Personal + Geographical Contacts	108.85	474	10

Table 5 shows the connectivity analysis of all nine types of contact. We can see that hospital admission history has the highest connectivity and hospital visit is the second highest. This is consistent with the fact that patients are primarily infected in hospitals. In personal contacts, family member has the highest connectivity. However, its average degree and number of components is still lower than any type of contact in the geographical contacts category.

Table 5. Connectivity Analysis for the Nine Types of Contacts

Main Category	Type of Contacts	Average Degree	Maximum Degree	Number of Components
Personal Contacts	Family Member	0.21	4	893
	Roommate	0.03	2	946
	Colleague/Classmate	0.06	3	934
	Close Contact	0.02	1	949
Geographical Contacts	Foreign Country Travel	2.73	41	848
	Hospital Visit	**10.08**	**105**	**753**
	High Risk Area Visit	1.39	36	924
	Hospital Admission History	**50.48**	**289**	**409**
	Workplace	4.69	61	823

5.2 Network Visualization

We use the contact network with all available personal and geographical contacts to explore the differences in network topology between the contact network and its transformed one-mode network. In this contact network, there are 961 patient nodes, 22 area nodes, 14 hospital nodes, and a total of 1313 edges. Figures 3 and 4 show the contact network and its transformed one-mode network. As seen in the one-mode network shown in Figure 4, with only patient nodes, it is difficult to distinguish the interconnections between them through the network visualization when their relationships get complicated. Compared to its one-mode network, the contact network clearly shows the role of those geographical locations in disease transmission. Since there were several hospital outbreaks during the SARS period in Taiwan, we can see from Figure 3 that the majority of patient nodes are spread around the hospital nodes in the contact network.

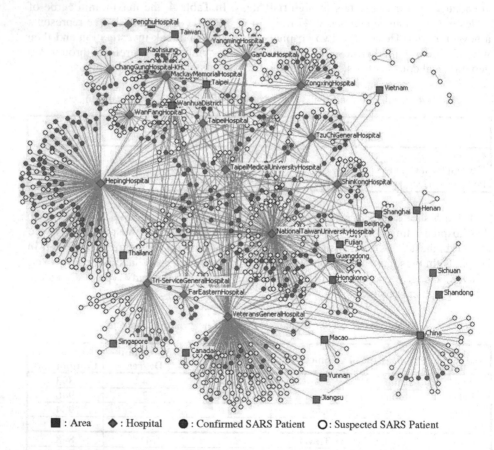

■ : Area   ◆ : Hospital   ● : Confirmed SARS Patient   O : Suspected SARS Patient

Fig. 3. Contact Network with All Available Personal and Geographical Contacts

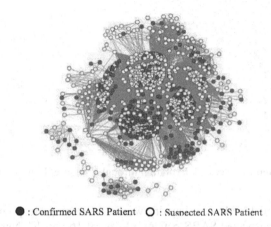

● : Confirmed SARS Patient   ○ : Suspected SARS Patient

Fig. 4. Transformed One-Mode Network with Only Patient Nodes

Furthermore, incorporating geographical locations into modeling of disease transmission also helps to reveal some potential people who act as bridges to transfer a disease from one subgroup to another one. Figure 5 shows the potential bridges among hospitals and households in the contact network.

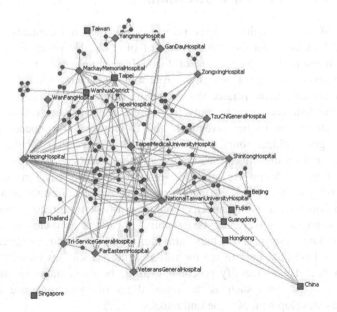

Fig. 5. Potential Bridges among Hospitals and Households in the Contact Network

For a hospital outbreak, including geographical contacts in the network is also useful to see the possible disease transmission scenario. Figure 6 demonstrates the evolution of a small contact network of Heping Hospital through the onset dates of

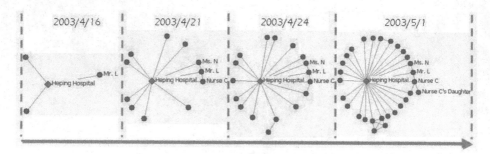

Fig. 6. Example of Network Evolution through the Onset Dates of Symptoms

symptoms. On April 16, Mr. L, a laundry worker in Heping Hospital, had a fever and was reported as a suspected SARS patient. On April 16 and 17, Nurse C took care of Mr. L. On April 21, Ms. N, another laundry worker, and Nurse C began to have symptoms. On April 24, Heping Hospital was reported to have a hospital outbreak. On May 1, Nurse C's daughter had a fever. From the evolution of the network, we can clearly see the development of the hospital outbreak.

6   Conclusions and Future Directions

The studies of SNA in epidemiology primarily use personal contacts to construct social networks and model the transmission of diseases. However, for the diseases which also transmit through the air or through contaminated inanimate objects, personal contacts alone are not enough to explain the underlying transmission context. Previous studies have demonstrated that including geographical contacts in SNA can reveal potential contacts among patients. In this research, by using Taiwan SARS data as the test dataset, we further investigate the differences in connectivity between personal and geographical contacts in the network construction for these diseases. From our research results, we can see that geographical contacts provide much higher connectivity in network construction than personal contacts. Therefore, for modeling the transmission of these diseases, incorporating geographical contacts into SNA is necessary to construct a well-connected contact network for investigation. We also explore the strengths of including geographical locations as nodes in the network visualization. Our results show that introducing geographical locations in SNA provides a good way not only to see the role that those locations play in the disease transmission but also to identify potential bridges between those locations. If we apply some context data, such as the onset dates of symptoms, we can further understand the development of some outbreaks.

For SARS, there is evidence showing that some patients were infected in flights or trains [12]. Incorporating transportation contacts into SNA will be our next focus. In addition, including geographical locations into SNA helps to reveal potential contacts but also brings some noise in network analysis. Therefore, we will also study approaches to filter those potential contacts and find the critical ones.

Acknowledgement

This work is supported by the National Science Foundation Digital Government Program through Grant #EIA-9983304, National Science Foundation Information Technology Research Program through Grant # IIS-0428241, and United States Department of Agriculture through Grant # 2006-39546-17579.

References

1. Abernethy, N.F.: Automating Social Network Models for Tuberculosis Contact Investigation. Ph.D. Dissertation. Department of Medicine, Stanford University, Stanford (2005)
2. Centers for Disease Control and Prevention (CDC): Severe Acute Respiratory Syndrome - Singapore, 2003. Morbidity and Mortality Weekly Report 52 (2003) 405-411
3. Chen, Y.C., Huang, L.M., Chan, C.C., Su, C.P., Chang, S.C., Chang, Y.Y., Chen, M.L., Hung, C.C., Chen, W.J., Lin, F.Y., Lee, Y.T.: SARS in Hospital Emergency Room. Emerging Infectious Diseases 10 (2004) 782-788
4. Ghani, A.C., Swinton, J., Garnett, G.P.: The Role of Sexual Partnership Networks in the Epidemiology of Gonorrhea. Sexually Transmitted Diseases 24 (1997) 45-56
5. Hsueh, P.R., Yang, P.C.: Severe Acute Respiratory Syndrome Epidemic in Taiwan, 2003. Journal of Microbiology, Immunology and Infection 38 (2005) 82-88
6. Klovdahl, A.S.: Social Networks and the Spread of Infectious Diseases: the AIDS Example. Social Science & Medicine 21 (1985) 1203-1216
7. Klovdahl, A.S., Graviss, E.A., Yaganehdoost, A., Ross, M.W., Wanger, A., Adams, G.J., Musser, J.M.: Networks and Tuberculosis: an Undetected Community Outbreak Involving Public Places. Social Science & Medicine 52 (2001) 681-694
8. Klovdahl, A.S., Potterat, J.J., Woodhouse, D.E., Muth, J.B., Muth, S.Q., Darrow, W.W.: Social networks and Infectious Disease: the Colorado Springs Study. Social Science & Medicine 38 (1994) 79-88
9. McElroy, P.D., Rothenberg, R.B., Varghese, R., Woodruff, R., Minns, G.O., Muth, S.Q., Lambert, L.A., Ridzon, R.: A Network-Informed Approach to Investigating a Tuberculosis Outbreak: Implications for Enhancing Contact Investigations. The International Journal of Tuberculosis and Lung Disease 7 (2003) S486-S493
10. Meyers, L.A., Pourbohloul, B., Newman, M.E.J., Skowronski, D.M., Brunham, R.C.: Network Theory and SARS: Predicting Outbreak Diversity. Journal of Theoretical Biology 232 (2005) 71-81
11. Morris, M.: Epidemiology and Social Networks: Modeling Structured Diffusion. Sociological Methods Research 22 (1993) 99-126
12. Peiris, J.S.M., Yuen, K.Y., Osterhaus, A.D.M.E., Stohr, K.: The Severe Acute Respiratory Syndrome. The New England Journal of Medicine 349 (2003) 2431-2441
13. Rothenberg, R.B., Narramore, J.: Commentary: the Relevance of Social Network Concepts to Sexually Transmitted Disease Control. Sexually Transmitted Diseases 23 (1996) 24-29
14. Rothenberg, R.B., Potterat, J.J., Woodhouse, D.E., Muth, S.Q., Darrow, W.W., Klovdahl, A.S.: Social Network Dynamics and HIV Transmission. AIDS 12 (1998) 1529-1536

15. Rothenberg, R.B., Sterk, C., Toomey, K.E., Potterat, J.J., Johnson, D., Schrader, M., Hatch, S.: Using Social Network and Ethnographic Tools to Evaluate Syphilis Transmission. Sexually Transmitted Diseases 25 (1998) 154-160
16. Shen, Z., Ning, F., Zhou, W., He, X., Lin, C., Chin, D.P., Zhu, Z., Schuchatt, A.: Superspreading SARS Events, Beijing, 2003. Emerging Infectious Diseases 10 (2004) 256-260
17. Yu, I.T.S., Li, Y., Wong, T.W., Tam, W., Chan, A.T., Lee, J.H.W., Leung, D.Y.C., Ho, T.: Evidence of Airborne Transmission of the Severe Acute Respiratory Syndrome Virus. The New England Journal of Medicine 350 (2004) 1731-1739

A Model for Characterizing Annual Flu Cases

Miriam Nuño and Marcello Pagano

Department of Biostatistics,
Harvard School of Public Health, Boston, MA 02115, USA
mnuno@hsph.harvard.edu, pagano@biostat.harvard.edu

Abstract. Influenza outbreaks occur seasonally and peak during winter season in temperate zones of the Northern and Southern hemisphere. The occurrence and recurrence of flu epidemics has been alluded to variability in mechanisms such temperature, climate, host contact and traveling patterns [4]. This work promotes a Gaussian–type regression model to study flu outbreak trends and predict new cases based on influenza–like–illness data for France (1985–2005). We show that the proposed models are appropriate descriptors of these outbreaks and can improve the surveillance of diseases such as flu. Our results show that limited data reduces our ability to predict unobserved cases. Based on laboratory surveillance data, we prototype each season according to the dominating virus (H3N2, H1N1, B) and show that high intensity outbreaks are correlated with early peak times. These findings are in accordance with the dynamics observed for influenza outbreaks in the US.

1   Background

Seasonal variation of infectious diseases is associated with several factors that include, environmental mechanisms, host–specific behavior and pathogen's ability to continuously invade the host population. The influenza virus, a well studied pathogen, is known for its ability to continuously invade the human host by constantly mutating and successfully evading a host's immune system. Due to constant minor (antigenic drift) and major changes (antigenic shift) in the virus surface proteins, flu vaccines are updated yearly to enhance protection against new infections. Influenza seasons occur during winter in temperate zones of the Northern and Southern hemisphere. It is estimated that some 50 million people get infected, more than 200,000 people are hospitalized from flu complications, and about 47,200 people die from flu each year.

A goal of research in bio–surveillance of communicable diseases such as influenza involves the development and implementation of reliable methods for early outbreak detection [1]. Effective surveillance methods enhance the preparedness and facilitate immediate response from health officials in the event of epidemic and pandemic events [3, 5, 8, 9]. The recent SARS outbreak (2002–2003) event showed that control and containment of the outbreak was mainly based on rapid diagnosis coupled with effective patient isolation according to the modeling in [2]. A commonly, and widely used method for automatic detection of

D. Zeng et al. (Eds.): BioSurveillance 2007, LNCS 4506, pp. 37–46, 2007.
© Springer-Verlag Berlin Heidelberg 2007

flu epidemics from time–series data was proposed by Serfling in 1963 [6]. Since then, the Center for Disease Control and Prevention has implemented the Serfling methodology to parameterize a baseline model based on statistical expectations (95% confidence interval of the baseline) by training data from non–epidemic years. A surveillance system based on the Serfling methodology signals an epidemic whenever the observed time–series data exceeds a threshold. The model assumes an average mortality (β_0), linear trend ($\beta_1 t$), and a 52–week cyclical period denoted by $\beta_2 cos(2\pi t/52) + \beta_3 sin(2\pi t/52)$. This model

$$Y(t) = \beta_0 + \beta_1 t + \beta_2 cos(2\pi t/52) + \beta_3 sin(2\pi t/52)$$

assumes that flu outbreaks are unimodal, cyclical and symmetric between the peak and troughs.

One of the aims of flu surveillance is the early detection of outbreaks, however, understanding the underlying mechanisms driving the observed fluctuations can be instrumental in developing effective monitoring systems. We study a time series regression model that summarizes outbreak trends and describes the observed seasonality. We apply the model to influenza–like–illness (ILI) weekly data for France reported during 1985–2005. We estimate the model parameters through least squares and validated the model numerically (adjusted R^2) and graphically (residual analysis). We assess the impact of timely reporting in predicting new flu cases. Finally, we study the correlation of outbreak peak times, intensity for virus-specified seasons and discuss the relevance of the proposed model for surveillance and prediction of flu outbreaks.

2 Methods

Weekly data of influenza–like–illness is illustrated in Figure 1 and demonstrates that flu outbreaks are highly seasonal with irregular intervals. The model propose incorporates several features that are characteristic of flu outbreaks. In particular, the model adjusts for variable peak times, intensity, and duration of outbreaks. Although the majority of flu outbreaks exhibit single peaks, Figure 1 shows that multiple peaks are also possible (Figure 1: the seasons 91–92, 97–98, 00-01 exhibit such behavior).

2.1 Statistical Model

We apply a Gaussian–type regression model to weekly influenza–like–illness (ILI) epidemic data to study outbreak trends for France during 1985–2005. For each year, we estimate the intensity, time of peak and duration of these outbreaks using least squares. The time series $Y_i(t)$ denotes the number of ILI cases observed in year i at the predictor time t. The value of j denotes the number of peaks considered in the model. That is, $j=1$ represents a single peak outbreak while $j=1,2$ assumes multiple peaks. The general form of the Gaussian model is as follows:

$$Y_i(t) = a_{i_0} e^{-b_{i_0} t} + \sum_{j=1}^{2} a_{ij} e^{-\left(\frac{t - b_{ij}}{c_{ij}}\right)^2}. \tag{1}$$

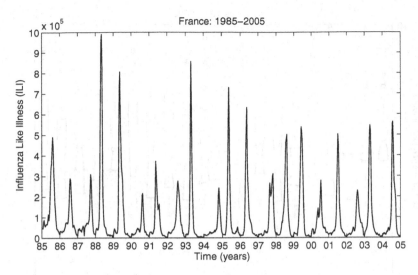

Fig. 1. Influenza–Like–Illness (ILI) weekly data for France (1985–2005)

Parameter a_{i_0} estimates the baseline (at time zero) of i–th year outbreak, a_{i_1} and a_{i_2} estimate the intensity of each peak in the i–th year, b_{i_0} estimate the decay rate for the background model, b_{i_1} and b_{i_2} evaluate the time of each peak, and c_{i_1} and c_{i_2} describe the duration of these outbreaks.

We identify outbreak trends for all seasons together, as well as, for outbreaks that are dominated by specific virus subtypes (H3N2, H1N1, B) by fitting the Gaussian model with single and multiple peaks. For each model, we estimate the mean, median and standard deviation of each of the parameters fitted in these models (e.g. outbreak peak time, intensity and duration). We summarize and compare the goodness of fit for each model numerically and graphically.

2.2 Dominating Virus Subtypes in Epidemic Seasons

Using laboratory surveillance data, influenza seasons (1985–2000) were summarized according to the prototype strain responsible each year. From the 20 seasons studied, virus A (H3N2) predominated 13 seasons, A (H1N1) dominated 3 seasons and the remaining 4 seasons are dominated by B type viruses. Figure 2 gives a box–plot description of the data for each season. The prototype seasons are distinguished by solid (H3N2), dotted (H1N1) and boxed (B) notches in order to illustrate the frequency and magnitude of these outbreaks and the corresponding dominating viruses.

2.3 Measuring the Uncertainty of Predicting New Cases

Based on the bimodal regression model, we estimate the likelihood of predicting ILI cases for unobserved times. We calculated prediction bounds for a new

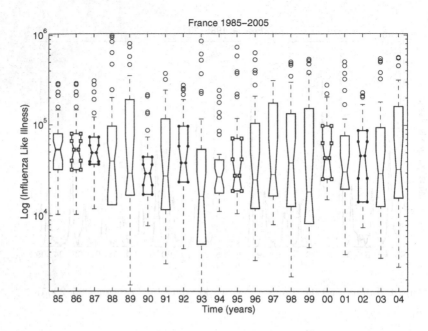

Fig. 2. French ILI data reported weekly. Solid, dotted and boxed notches denote outbreaks dominated by H3N2, H1N1 and B virus subtypes, respectively.

Table 1. Parameter estimation for single and bimodal peak models based on French ILI data from 1985 to 2005. We calculate the mean, median and standard deviation for each of the parameters estimated. Note that a_1 describes the model baseline, a_2 and a_3 the intensity of peaks 1 and 2, with peak times at b_2 and b_3, and corresponding duration given by c_1 and c_2. Parameter b_1 denotes the decay rate of each season. Non applicable findings are denoted by na.

$Y_i(t)$	a_1	a_2	b_1	b_2	c_1	a_3	b_3	c_2	R^2	adj R^2	RMSE
1985–2005											
Model 1											
mean	30464	507823	0.20	16.2	3.3	na	na	na	0.9550	0.9485	29092
median	26640	497000	0.03	14.9	3.2	na	na	na	0.9728	0.9689	25990
STD	20419	208105	0.52	5.4	0.9	na	na	na	0.0471	0.0538	17561
Model 2											
mean	29864	383466	0.01	13.6	3.3	240189	17.8	3.8	0.9919	0.9899	13133
median	27370	413500	0.04	11.5	2.0	240300	17.1	3.5	0.9934	0.9915	11780
STD	16188	262955	0.08	4.8	1.7	161857	4.9	1.9	0.0062	0.0077	4077

observation assuming that data was not available (extrapolation). The prediction bounds are calculated simultaneously and measure the confidence that a new observation lies within the interval regardless of time.

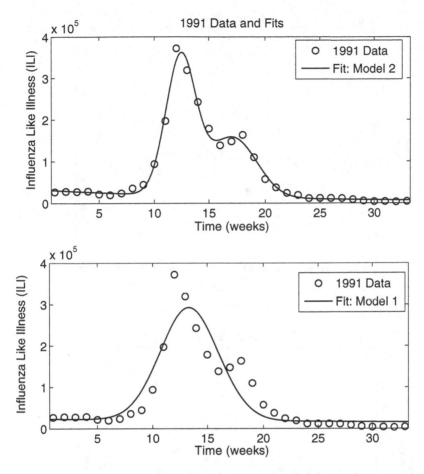

Fig. 3. Fits obtained with regression Models 1 and 2

3 Results

To assess the outbreak trends for all seasons of data available, we fit Models 1 and 2 to each year of data. We estimate the parameters of the single peak model (denoted by Model 1) and multiple peak model (denoted by Model 2) through least squares for each of the 20 years of data available (Table 1). After estimating the best fits for all 20 years, we calculate the mean, median and standard deviation (STD) of the parameters in these models. Fitting results of Model 1 yield a mean of 4.3×10^5, 4.4×10^5 median and 2.2×10^5 STD for the parameter estimating the intensity of the outbreaks. Similarly, we estimate the mean, median and standard deviation for the peak times and obtain a 18.4 weeks, 17.2 weeks, and 5 weeks, respectively. Table 1 shows that Model 2 describes the data better than Model 1 (adjusted R^2). We illustrate the actual fit of both Model for the 1991 season. These fits illustrate that the single peak model is ill specified to capture the bimodality of the data.

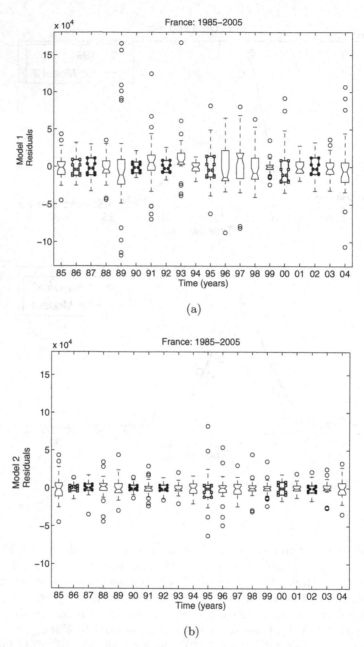

Fig. 4. Residual plots for Model 1 (a) and Model 2 (b) for French ILI data from 1985-2005. Solid, dotted and boxed notches denote the residuals corresponding to H3N2, H1N1 and B type strains.

We find strong evidence that Model 2 fits the data better than Model 1. The goodness of fit of each of these models was assessed by analyzing their residuals

for each of the years fitted. We illustrate the goodness of fit of these model for
the 1990-1991 flu season. Figure 3 illustrates the fits obtained with Model 2
(top–panel), Model 1 (bottom–panel) for these fits.

In order to investigate the goodness of fit of Model 1, we calculated these
statistics for the adjusted R^2 values of this data. The adjusted R^2 values estimated were 0.9423 (mean), 0.9606 (median) and 0.0588 (STD). In order to
compare the goodness of fit of Models 1 and 2, we carry out a similar analysis
for Model 2 (see Table 1). Our results show that Model 2 improves the data
fit. That is, the adjusted R^2 estimates for the mean is improved from 0.9423
to 0.987, median from 0.9609 to 0.988 with corresponding standard deviation
reduced by approximately 85% (from 0.0588 to 0.009). We further assess these
fits graphically by calculating the residuals for each model. Figure 4 (top–panel)
illustrates the residuals for Model 1 and Figure 4 (bottom–panel) for Model 2.

3.1 Correlating Trends with Subtype–Specific Outbreaks

We further assessed the trends of these outbreaks by analyzing them according
to dominating virus in each season. Our results show that intensity of H3N2
outbreaks were significantly higher than H1N1 and B subtypes combined ($p =$
0.0226, Kruskal-Wallis test, two-tailed). Table 2 also shows that H3N2 outbreaks
tend to peak sooner than H1N1 and B. We carried out a similar test to assess
any significant difference among the peak times for H3N2 and those for H1N1
and B and find that peak times for H3N2 subtype outbreaks occur earlier than
H1N1 and B. Note that these results are supported by the parameter estimates
obtained from fitting Model 1 and Model 2. For each subtype dominant season,
our goodness of fit results (adjusted R^2) showed that the latter model improves
the fit.

Although the data in this study was available weekly, we assumed several scenarios with limited data and assessed the likelihood of predicting new
ILI cases for unspecified times. We assumed that data is available weekly (as
in the current study), biweekly and monthly. Figure 5 illustrates the prediction bounds (dashed–dotted) obtained assuming weekly (left–panel), biweekly
(middle–panel) and monthly (left–panel) data. Evaluating the prediction bounds
for each of the scenarios shows that our ability to predict new ILI cases decreases
as less data becomes available. That is, we show that for weekly available data we
can predict all data points with high certainty since all data lies within the 95%
confidence interval of prediction. As the data becomes more scarce (biweekly),
we show that we are no longer able to predict the highest intensity data of the
outbreak (Figure 5: middle–panel) . Finally, for monthly available data, we show
that we are no longer able to predict the bimodality of our data in addition to
the highest intensity peak.

4   Discussion

The findings in this study promote two Gaussian–type of regression models that
assess influenza outbreak trends. Unlike the well–known cyclical Serfling model,

Table 2. Fit results implementing Model 1 and Model 2 for ILI data that is grouped according to the subtype–specific strains (H1N1, H3N2, and B) dominating in each season. Note that a_1 describes the model baseline, a_2 and a_3 the intensity of peaks 1 and 2, with peak times at b_2 and b_3, and corresponding duration given by c_1 and c_2. Parameter b_1 denotes the decay rate of each season. Non applicable findings are denoted by na.

$Y_i(t)$	a_1	a_2	b_1	b_2	c_1	a_3	b_3	c_2	R^2	adj R^2	RMSE
H3N2											
Model 1											
mean	30464	507823	0.20	16.2	3.3	na	na	na	0.9550	0.9485	29092
median	26640	497000	0.03	14.9	3.2	na	na	na	0.9728	0.9689	25990
STD	20419	208105	0.52	5.4	0.9	na	na	na	0.0471	0.0538	17561
Model 2											
mean	29864	383466	0.01	13.6	3.3	240189	17.8	3.8	0.9919	0.9899	13133
median	27370	413500	0.04	11.5	2.0	240300	17.1	3.5	0.9934	0.9915	11780
STD	16188	262955	0.08	4.8	1.7	161857	4.9	1.9	0.0062	0.0077	4077
H1N1											
Model 1											
mean	42843	364367	0.01	16.3	2.5	na	na	na	0.9133	0.9011	24683
median	45400	246100	0.01	17.9	2.5	na	na	na	0.9559	0.9497	26710
STD	12404	240319	0.01	4.1	0.4	na	na	na	0.0945	0.1080	7091
Model 2											
mean	33040	318300	0.11	14.0	2.3	115903	21.7	4.9	0.9825	0.9775	15410
median	33600	212800	0.01	12.6	2.4	84370	19.5	2.1	0.9802	0.9746	11870
STD	2644	285561	0.19	3.4	0.1	104945	5.2	5.64	0.0101	0.0130	10785
B											
Model 1											
mean	31985	221450	0.004	21.1	3.1	na	na	na	0.9588	0.9528	14030
median	29930	216950	0.01	20.5	2.9	na	na	na	0.9582	0.9521	13950
STD	13091	35613	0.02	2.5	0.9	na	na	na	0.0205	0.0234	3692
Model 2											
mean	28325	78820	0.04	18.0	5.7	177523	21.7	3.8	0.9867	0.9830	8578
median	24795	72175	0.05	19.6	5.6	187100	21.4	2.7	0.9865	0.9827	8776
STD	13293	49331	0.03	3.9	3.2	66625	2.14	2.6	0.0050	0.0064	1764

these models adjust for variability in time of peaks, intensity and duration of outbreaks. We show that these models are highly effective in describing outbreak trends, and thereby facilitate the assessment of flu patterns. The data presented in this study illustrates that flu outbreaks depict multiple peaks and therefore appropriate models are needed to regard for these dynamics. A residual evaluation of the fits of these models shows that these models are highly effective in describing the data presented here. These models show that they are highly effective in describing the data presented here, particularly, the bimodal model.

The results of this study support previous observations of the correlation in the time of peak and intensity of influenza epidemics for A (H3N2) virus outbreaks for the US [7]. That is, high intensity outbreaks tend to occur early–on the season, while lower intensity outbreaks (H1N1 and B) occur later. Moreover, our study

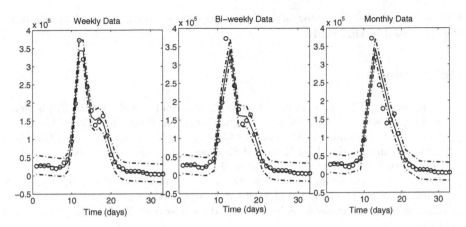

Fig. 5. Predictions based on weekly (left–panel), biweekly (middle–panel) and monthly data available (right–panel)

suggests that low–intensity peaks are more likely to be followed by high–intensity peaks since it is possible that frequently dominating viruses benefit from the recruitment of susceptibles during for almost two years. However, less dominating virus subtypes (H1N1 and B) do not rely on this recruiting advantage, and therefore face a continuous challenge to become established in the population as the dominating virus.

Effective surveillance of infectious diseases involves the timely assessment and implementation of monitoring systems aimed to reduce morbidity and mortality while minimizing societal disruption. Influenza surveillance is a combined effort of virological identification, clinical and epidemiological monitoring of multiple source data such as influenza–like–illness, pneumonia and influenza related mortality, hospitalization, to name a few. However, in this study we show that characterizing outbreak trends improves our understanding of the underlying mechanisms driving influenza epidemics, and therefore, is key for developing effective surveillance systems.

A critical limitation of this work lies in the prediction of unobserved cases. Our work evaluates the likelihood of predicting ILI cases during a particular season for unobserved times (retrospectively), however, we do not assess prediction of cases in future seasons. It is evident that effective surveillance systems should include a clear understanding of outbreak trends and retrospective assessment of unobserved cases. Moreover, the prediction of future cases based on current and historical data is an essential component of effective surveillance. To this end, our current research efforts are placed in predicting flu cases for future seasons based on the descriptive models proposed herein.

Acknowledgments. Miriam Nuño was funded by the Yerby Postdoctoral Fellowship and the National Institutes of Health grant T32AI07358. The research of Marcello Pagano was funded by the National Institutes of Health grant R01EB006195. The authors are thankful to Martine Valette of NIC–Lyon

Laboratoire de Virologie for providing prototype strain isolate information for the data used in this study.

References

[1] Buffington, J., Chapman, L. E., Schmeltz, L. M., Kendal, A, P.: Do family physician make good sentinels for influenza?, Arch. Fam. Med. **2:8** (1993) 859–864.
[2] Chowell, G., Fenimore, P, W., Castillo–Garsow, M, A., Castillo–Chavez, C.: SARS outbreaks in Ontario, Hong Kong and Singapore: the role of diagnosis and isolation as a control mechanism, J. Theor. Biol. **24** (2003) 1-8.
[3] Lazarus, R., Kleinman, K, P., Dashevsky, I., DeMaria, A., Platt, R.: Using automated medical records for rapid identification of illness syndromes (syndromic surveillance): the example of lower respiratory infection, BMC Public Health **1:9** (2001).
[4] Lofgre, E., Fefferman, N., Naumov, Y, N., Gorski, J., Naumova, E, N.: Infuenza Seasonality: Underlying Causes and Modeling Theories, J. Virol. (2006) 1680–1686.
[5] Reis, B,Y., Mandl, K,D.: Integrating syndromic surveillance data accross multiple locations: effects on outbreak detection performance, AMIA Annu. Symp. Proc. (2003) 549–53.
[6] Serfling, R, E.: Methods for current statistical analysis of excess Pneumonia-Influenza deaths, Public Health Reports **78:6** (1963) 494–506.
[7] Simonsen, L., Reichert, T, A., Viboud, C., Blackwelder, W, C., Taylor, R, J., Miller, M. A.: Impact of Influenza Vaccination on Seasonal Mortality in the US Elderly Population, Arch. Intern. Med. **165** (2005) 265–272.
[8] Toubiana, L., Flahault, A.: A space-time criterion for early detection of epidemics of influenza-like-illness, Eur. J. Epidemiol. **14:5** (1998) 465–70.
[9] Upshur, R, E., Knight, K., Goel, V.: Time-series analysis of the relation between influenza virus and hospital admissions of the elderly in Ontario, Canada, for pneumonia, chonic lung disease, and congestive heart failure, Am. J. Epidemiol. **149:1** (1999) 85–92.

Population Dynamics in the Elderly: The Need for Age-Adjustment in National BioSurveillance Systems

Steven A. Cohen[1] and Elena N. Naumova[2]

[1] Department of Population, Family, and Reproductive Health, Johns Hopkins Bloomberg
School of Public Health, Baltimore, Maryland
scohen@jhsph.edu
[2] Department of Public Health and Family Medicine, Tufts University School of Medicine,
Boston, Massachusetts
Elena.Naumova@tufts.edu

Abstract. With the growing threat of pandemic influenza, efforts to improve
national surveillance to better predict and prevent this disease from affecting the
most vulnerable populations are being undertaken. This paper examines the
utility of Medicare data to obtain age-specific influenza hospitalization rates for
historical analyses. We present a novel approach to describing and analyzing
age-specific patterns of hospitalizations using Medicare data and show the
implications of a dynamic population age distribution on hospitalization rates.
We use these techniques to highlight the utility of implementing a real-time
nationwide surveillance system for influenza cases and vaccination, and discuss
opportunities to improve the existing system to inform policy and reduce the
burden of influenza nationwide.

Keywords: real-time surveillance, influenza, age-adjustment, elderly, Medicare.

1 Introduction

Influenza is a significant public health problem in the United States. From 1968 to
1998, annually, there were, on average, over 6,000 deaths that were directly
attributable to influenza [11], and over 36,000 annual deaths considered to be
pneumonia, induced by influenza, or influenza-related (P&I) [13]. Nationwide, there
are over one million hospitalizations in the population age 65 and over for P&I-
related conditions annually [14]. An influenza pandemic could result in serious
economic consequences, as well. In the US alone, a pandemic causing infection in 15-
35% of the population could result in nearly $170 billion in direct and indirect
costs [5].

The elderly are particularly vulnerable to the morbidity and mortality associated
with influenza. The age-specific mortality rate from P&I deaths is nearly 100 times
higher in people aged 65 and over (22.1/100,000 person-years) than for children
under 1 year of age, the group with the second-highest influenza mortality rate
(0.3/100,000 person-years). From 1990 through 1998, 90% of influenza-associated
deaths occurred among the population age 65 and older [13].

Accurate and timely surveillance of influenza at the nationwide level is
challenging. In the US, influenza surveillance is accomplished primarily through

D. Zeng et al. (Eds.): BioSurveillance 2007, LNCS 4506, pp. 47–58, 2007.

reporting of laboratory-confirmed cases to the Centers for Disease Control and Prevention (CDC). As part of the influenza surveillance effort, CDC also maintains a specialized mortality and morbidity reporting system, the 121 Cities Mortality Reporting System, where influenza deaths are reported from 122 cities and metropolitan areas throughout the US within two to three weeks of the death. These surveillance issues become even more important when considering that this disease is largely preventable [4].

This report examines the Center for Medicare and Medicaid Services (CMS) surveillance data on influenza-associated hospitalizations. We describe the utility and accuracy of using smoothed data-based, age-specific P&I hospitalization rate model parameters as an outcome measurement of P&I morbidity. We also examine the effect of population dynamics, namely, the rapidly changing age distributions in the elderly population, on estimating rates of age-specific P&I hospitalizations. This case study will underscore the shortcomings in the current national surveillance systems for influenza and highlight some potential opportunities for improvement.

2 Methods

2.1 Data

The data for this analysis come from two sources: CMS and the US Census Bureau. The CMS dataset contains all hospitalization records from 1998 through 2002. The populations served by Medicare include the general population age 65 and over, along with those with end-stage renal disease. The Medicare database covers 98% of the US population age 65 and above [7]. The US Census data used in this analysis consisted of the state-level age-specific population by state from Census 2000, abstracted from Summary File 1.

2.2 Data Abstraction

We abstracted approximately 6.2 million records containing all P&I hospitalizations (ICD codes 480-487) for the Medicare population age 65 to 99. We excluded cases from the population age 100 and over. The cases extracted occurred between July 1998 and June 2002 by single year of age, state of primary residence, and influenza year. An influenza year was defined to start on July 1 and end on June 30 of the following year. We examined four consecutive influenza years in this analysis (1998-99 through 2001-02).

To obtain the denominator of the age-specific P&I hospitalization rate, we gathered data from the 2000 US Census containing the population distribution of each state by single year of age for April 1, 2000. The total count of P&I cases by age, state, and season was then divided by the Census 2000 population for that state and age to obtain P&I hospitalization rates for each influenza season.

2.3 Estimation of P&I Hospitalization Parameters

Because the morbidity rates for P&I morbidity tend to increase exponentially with age in the elderly population [2], the age-specific rates were modeled against age according to the model

$$\log(\text{rate}_{ij}) = \beta_{0i} + \beta_{1i}(\text{age}_j-65) + \varepsilon_i, \text{ where } i = \text{state}, j = \text{age} \qquad (1)$$

Thus, β_{0i} represents the estimated intercept for the state, which represents the log of the estimated P&I hospitalization rate in each state at age 65. β_{1i} is the rate of change in the log-transformed P&I hospitalization rates by age for each state. The larger the value of β_{1i} was, the sharper the increase in P&I hospitalization rates by age in the state was. These model parameters comprised the two main variables of analysis.

2.4 Statistical Analysis

Validity of Outcome Measures. To ensure that the outcome measures, the state-specific model parameters for the P&I hospitalization rates regressed against age, provide reliable estimates of the true rates, we obtained R-squared statistics for each of the regression models for each state and for each influenza season.

Sensitivity and Adjustment for Population Change. Because the age-specific population used in the denominator of the P&I hospitalization rates changes over time, we explored the relationship between using a dynamic denominator, based on US Census Bureau estimates for the population in each year (1998-2002) and using the static measures from Census 2000. The results of changing this denominator are illustrated using a series of Lexis diagrams with contour plots of population change. Lexis contour maps are derived from Lexis diagrams, which are graphical tools commonly used in demographic research. Lexis contour maps have calendar time, often expressed in years, on the horizontal axis and age on the vertical axis. The contours can represent any variable in which one seeks to analyze age and time effects simultaneously.

Software. SAS version 9.0 was used for all statistical analyses. For graphs, Microsoft Excel 2003 was used, along with S-PLUS version 7.0 for Lexis contour maps.

3 Results

3.1 Validity of Outcome Measures

The state-level age-specific P&I rates were regressed against age to obtain the outcome measures, the slope and intercept of the models by state. To assess the model fit, we obtained R-squared values for each state and season (Table 1).

In general, this approach to parameterizing the state-level age-specific P&I mortality rates in the older population was fairly precise. In 145 of the 204 cases (71.1%), the R-squared value was above 0.90, verifying that P&I hospitalization rates increase nearly exponentially with age from age 65 to age 99. These exponential models tended to perform less accurately in some states, such as Hawaii, where in the 1999-2000 season, the model R-squared was only 0.503. Moderate R-square values were seen in states such as Wyoming (0.769-0.901), Montana (0.723-0.942), Alaska (0.555-0.785), and others.

Table 1. R-squared values of state-level regression models by season

State	Season				State	Season			
	1998-1999	1999-2000	2000-2001	2001-2002		1998-1999	1999-2000	2000-2001	2001-2002
Alabama	0.976	0.968	0.966	0.984	Montana	0.782	0.942	0.801	0.723
Alaska	0.740	0.555	0.785	0.765	Nebraska	0.920	0.918	0.930	0.971
Arizona	0.862	0.938	0.936	0.916	Nevada	0.895	0.968	0.854	0.705
Arkansas	0.965	0.955	0.910	0.952	New Hampshire	0.876	0.921	0.819	0.741
California	0.983	0.991	0.990	0.957	New Jersey	0.982	0.979	0.965	0.933
Colorado	0.963	0.897	0.955	0.936	New Mexico	0.961	0.886	0.956	0.898
Connecticut	0.943	0.974	0.945	0.918	New York	0.987	0.975	0.974	0.975
Delaware	0.791	0.826	0.756	0.811	North Carolina	0.955	0.945	0.965	0.973
DC	0.852	0.873	0.828	0.875	North Dakota	0.946	0.889	0.961	0.951
Florida	0.966	0.990	0.970	0.983	Ohio	0.963	0.973	0.966	0.948
Georgia	0.972	0.961	0.977	0.963	Oklahoma	0.934	0.907	0.991	0.984
Hawaii	0.814	0.503	0.786	0.797	Oregon	0.880	0.897	0.923	0.908
Idaho	0.727	0.900	0.828	0.873	Pennsylvania	0.959	0.961	0.933	0.954
Illinois	0.977	0.969	0.988	0.972	Rhode Island	0.879	0.899	0.865	0.912
Indiana	0.992	0.959	0.966	0.949	South Carolina	0.949	0.942	0.937	0.966
Iowa	0.933	0.973	0.974	0.952	South Dakota	0.926	0.939	0.935	0.967
Kansas	0.943	0.959	0.967	0.940	Tennessee	0.983	0.973	0.985	0.969
Kentucky	0.975	0.963	0.974	0.978	Texas	0.977	0.960	0.976	0.979
Louisiana	0.986	0.975	0.896	0.960	Utah	0.887	0.754	0.881	0.903
Maine	0.843	0.895	0.924	0.954	Vermont	0.895	0.835	0.887	0.900
Maryland	0.968	0.950	0.879	0.948	Virginia	0.968	0.957	0.938	0.937
Massachusetts	0.974	0.927	0.960	0.957	Washington	0.978	0.979	0.887	0.940
Michigan	0.978	0.920	0.955	0.937	West Virginia	0.929	0.949	0.931	0.949
Minnesota	0.959	0.945	0.968	0.957	Wisconsin	0.927	0.966	0.927	0.830
Mississippi	0.923	0.988	0.971	0.955	Wyoming	0.769	0.901	0.821	0.775
Missouri	0.966	0.975	0.945	0.979					

Figure 1 shows a comparison of two states where the model performed markedly differently: Hawaii in the 1999-2000 season ($R^2 = 0.503$) and Indiana in 1998-1999 ($R^2 = 0.992$). This example suggests that much of the deviation from the predicted P&I rates for Hawaii comes from the population in the oldest age groups. The predicted rates are similar to the actual rates for ages 65 through the late 80's. From the late 80's through age 99, the actual rates show more random variation from year to year. This is most likely due to extremely small age-specific populations in this age group in Hawaii.

The same holds true for other states, as well. In contrast, the actual and predicted P&I rates are much more similar in Indiana throughout all age groups. Thus, one advantage of this technique is to smooth P&I hospitalization rates in cases such as Hawaii where the actual rates may have extra variation due to small population size leading to more noise in the estimates.

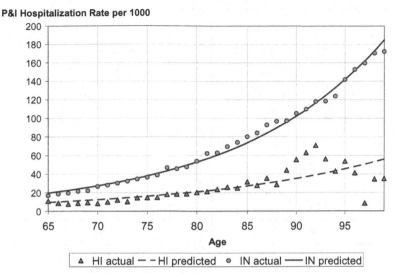

P&I Hospitalization Rate per 1000

Fig. 1. Actual and model-based predicted P&I rates by age for Hawaii (1999-2000) and Indiana (1998-1999)

To simplify the estimation procedure, we log-transformed the P&I rates as described above and multiplied those parameters by 100. The results of the models are shown as descriptive statistics in Table 2.

Table 2. Descriptive statistics for derived intercept and slope outcome variables

	Minimum	Maximum	Mean	Std. Dev.
Intercept 98-99	653.2	779.7	720.1	30.7
Slope 98-99	6.79	8.96	7.79	0.59
Intercept 99-00	654.0	777.2	720.5	31.6
Slope 99-00	5.34	9.16	7.87	0.71
Intercept 00-01	648.4	777.2	715.2	30.9
Slope 00-01	7.00	9.31	8.01	0.59
Intercept 01-02	659.9	785.8	723.4	29.5
Slope 01-02	6.96	9.29	8.01	0.56

This table shows that the intercept and slope parameters remained fairly steady over the time period of analysis.

Figure 2 illustrates the relationship between slope and intercept by state for each of the four seasons examined.

The slopes and intercepts in each state were inversely related to each other. States with higher intercepts tended to have lower slopes. This means that, typically, a state that had higher rates of P&I hospitalization at the younger ages had a lower subsequent rate of increase in P&I hospitalizations with age.

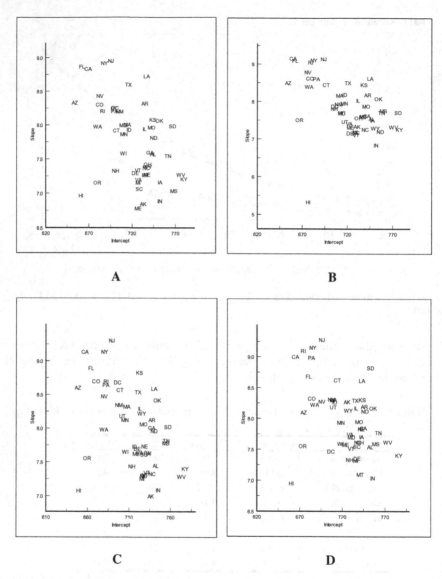

Fig. 2. Scatterplots of the relationship between state-specific slopes and intercepts for 1998-1999 (Panel A), 1999-2000 (Panel B), 2000-2001 (Panel C), and 2001-2002 (Panel D)

3.2 Adjustment for Population Change

The age-specific population estimates extracted from the US Census Population Estimates Branch for 2000-2005 are shown in Figure 3 for two states for ages 65-84. These panels show two examples of distinct patterns of population change on the state level. In Nevada, there was rapid population growth in the entire population age 65-84 over time. For example, there was a 25% increase in the population age 65 in Nevada between 2000 and 2005, and a 31% increase in the same time period in the population age 80. Rhode Island (Panel B), however, experienced population decline in the younger age brackets (67-79), but had a growing population in the oldest age groups (80-84). The Rhode Island population declined by 17% between 2000 and 2005, while the population age 84 increased by 15% over that same five-year span.

These have important implications for the calculation of P&I hospitalization rates. For example, in Nevada, where population growth is among the highest in the country, using Census 2000 populations in the denominator of rates would have a noticeable impact on the rates of disease. Figure 4 shows the percent difference between the age-specific P&I hospitalization rate using Census 2000 and Census estimates for individual years for the years 2000-2003. This graph shows that if Census 2000 is used instead of Census estimates for years other than 2000, the discrepancies between P&I rates generally grow with time.

The color gradient depicts the change in population as the following: yellow to red colors represent increases in age-specific populations, while green regions represent decreases in age-specific populations compared to Census 2000. Most of the graph shows yellow and red regions, which represent ages and years in which the P&I rate would be overestimated if using Census 2000 because the population has grown since the 2000 Census. There is a small section of the graph that appears green. There was a small population decline from 2000 to 2003 at age 72. If Census 2000 figures were used to calculate P&I hospitalization rates for 72-year-olds in 2003, this figure would actually represent a minor overestimate, since the population has decreased.

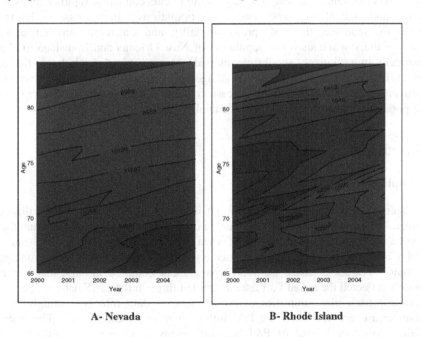

A- Nevada   **B- Rhode Island**

Fig. 3. Lexis contour plots of age-specific population for two states over time

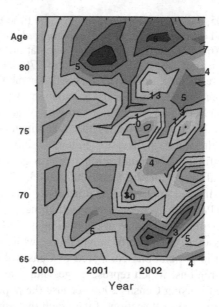

Fig. 4. Percent difference in P&I hospitalization rates by age and year for Nevada comparing Census 2000 and individual year Census estimates as the denominator

Some recent events illustrate that large-scale events can cause rapid changes to the structure and size of city- and state-level populations. In the case of Hurricane Katrina, for instance, the widespread mortality and out-migration that occurred because of the storm caused the population of New Orleans and Louisiana to change dramatically in a relatively small time interval. Migration and death due to the storm did not affect the population equally at all ages [15]. These changes can have serious consequences in the estimation of disease incidence that would not be captured if using population data only from the Decennial Census.

4 Discussion

4.1 Implications of Analysis

The results of this analysis have important implications for national surveillance of influenza. In terms of analytical techniques, a useful means of assessing the cumulative, season-based P&I hospitalization rates when single-year age-specific P&I rates are available is the method we employed, based on the regression of log P&I rate on age for each state. For most states, the model parameters obtained through this technique accurately reflected the actual P&I rates abstracted through the CMS data.

This approach also simplifies a vast amount of data into two straightforward, relevant characteristics about the P&I influenza rates in each state. The intercept provides the general level of P&I hospitalizations in each state, while the slope provides information on the most vulnerable section of the population, the oldest old. The slope, arguably, is the more important measure for several reasons. The actual age-specific P&I hospitalization rates are orders of magnitude higher in the oldest ages compared to the younger elderly. This means that small differences in slopes between states could potentially translate to large differences in the rates of P&I hospitalization in the oldest elderly population. One may challenge this notion by stating that although the P&I hospitalization rates are highest in the oldest age groups, the actual numbers of elderly in those age groups affected by P&I morbidity is low, simply because of smaller age-specific population sizes in the oldest elderly population. However, given that the majority of influenza-related hospitalizations and deaths occur in the older age categories, monitoring and modeling P&I hospitalization rates on this vulnerable population is particularly critical to prevent morbidity and delay mortality.

This method also adjusts for differences in the population composition between states. This is because the method is based on rates, as opposed to counts of cases, and those rates are specific to the state population age distribution. This method could also be applied to smaller areas of observation, such as counties, although the models would be expected to be less accurate due to smaller, less stable populations.

An important consideration when using this method, as with any other methods involving rates of disease, is the selection of the population as the denominator. As observed in the above results, small changes in population size have a noticeable effect on infection rates, especially considering the small population size in the older age groups.

4.2 Example: Establishing a Vaccination Surveillance System

A key function of the surveillance system is its utility in the monitoring and prevention of disease in the population. In regard to influenza prevention, the current national disease surveillance system does not contain information on the most widespread prevention mechanism: vaccination.

According to the Centers for Disease Control and Prevention (CDC), the primary means of controlling the detrimental effects of influenza is through annual vaccination [12]. The CDC currently recommends that certain population groups be vaccinated on an annual basis. Given an insufficient vaccine supply necessary for universal vaccination, the CDC recommends immunizing the elderly against influenza, and then people age 2 to 64 with comorbid conditions and who would be more susceptible than others in this age group, then the elderly who have comorbid conditions. Fourth on the list of priority groups are children age 6 to 23 months.

Although vaccinating the elderly against influenza has been shown to reduce P&I morbidity and mortality [8], recent research has questioned the policy of vaccinating the elderly for several reasons. First, vaccine efficacy in the elderly is low compared to other population subgroups [3]. Second, there is mounting evidence that herd immunity may form in the elderly as a result of vaccinating other population subgroups in which vaccine efficacy is high [9], [10]. Future research could help elucidate if an effective strategy to control influenza-related outcomes would be to vaccinate children, and if certain population-level factors modify morbidity and mortality in the elderly population [6], [16].

This research can be accomplished by improving the current national surveillance system for influenza. One way of improving national surveillance efforts is to implement a national surveillance system for influenza vaccination. Currently, influenza vaccination coverage in children, adults, and the elderly is conducted only through surveys, such as the National Immunization Survey and the Behavioral Risk Factor Surveillance System. As with any survey, the results are susceptible to biases, such as selection bias, and are associated with a level of uncertainty because the survey represents just a sample of the population. Another issue pertains specifically to the National Immunization Survey, which provides data on child immunizations. This survey provides data only on children age 19 to 35 months for influenza vaccinations. This excludes children age 3 years and older, which represents a large segment of the child population. This segment of the population could potentially be as or more important than the population age 19 to 35 months in transmitting influenza virus for several reasons. First, the population size of 3-17-year-olds is substantially larger than the population age 19 to 35 months in each state. Second, and perhaps more importantly, this population subgroup may potentially have a greater role in transmitting influenza virus both within their own population subgroup, as well as to other population subgroups, compared with the population 19 to 35 months of age. The influenza virus acquired in these settings could then be transmitted to other population subgroups through family contacts and in other public settings [7]. Thus, having a nationwide surveillance system that includes age-specific influenza vaccination coverage for all ages could further this research and help elucidate which population subgroups to vaccinate against influenza in order to achieve the maximum benefit in terms of morbidity and mortality reduction.

Timing is another key feature of a proposed national surveillance system for influenza vaccination. Theoretically, people are vaccinated before the influenza season begins, which typically occurs in the fall. However, the current vaccination surveys typically ask whether a person has been vaccinated against influenza in the past calendar year. If the period of interest is, for example, January through December of a given year, the vaccinations that are recorded in this survey may be from late in the prior season or in the fall of the current season, which would tend to push any potential associations between vaccination and indirect benefits toward the null. A nationwide surveillance system in which the timing of vaccination is recorded would reduce this contamination, and afford the opportunity to determine if the timing of vaccination is related to the timing and amplitude of the seasonal peak in P&I hospitalizations by population subgroups.

4.3 Opportunities to Improve National Influenza Surveillance

Medicare data on P&I hospitalizations can be used to perform a historical analysis, such as the one presented in this paper. However useful from an analytical perspective, Medicare data does not provide information in real-time. From an analytical perspective, this CMS data contains information on the entire Medicare-eligible population, as opposed to a sample of the elderly population. One can extract and model the timing, amplitude, and other features of the P&I hospitalization rates by state over time, and, in conjunction with improved immunization data quality, could estimate the relationships between immunization timing and intensity with P&I hospitalization timing and intensity.

Since Medicare data is not provided in real-time, it has little utility for measuring the burden of influenza in the current influenza season. This data also covers only one segment of the population. The best source of data on real-time influenza cases from the entire population is from the CDC's 121 Cities Mortality Reporting System. This provides up-to-date information on the latest reported influenza cases in the 122 cities and metropolitan areas associated with this reporting system. This system, however, does not include all of the rural areas of the country. According to a recent report by RAND [1], hospitalizations for acute infectious diseases actually occur more frequently in rural areas than in urban areas, suggesting that a simple extrapolation of the 121 Cities Mortality Reporting System to the national level would underestimate the true burden of influenza morbidity in the general population, especially in rural states.

5 Conclusion

This is just some of the rationale to underscore the importance of establishing and maintaining a real-time nationwide surveillance system for influenza. Combining the precision and coverage of the CMS data along with the timeliness of the 121 Cities Mortality Reporting System, a real-time surveillance system would be beneficial, not only for researchers, but for policymakers and other stakeholders involved with influenza prevention and control. This becomes especially critical, given the present threat of a worldwide influenza pandemic in the near future.

References

1. Farley, D.O., Shugarman, L.R., Taylor, P., Inkelas, M., Ashwood, J.S., Zeng, F., Harris, K.M.: Trends in Special Medicare Payments and Service Utilization for Rural Areas in the 1990s. RAND. Published June 2002
2. Horiuchi, S., Finch, C. E., Mesle, F., Vallin, J.: Differential Patterns of Age-related Mortality Increase in Middle Age and Old Age. J. Gerontol. A Biol. Sci. Med. Sci. 58 (2003) 495-507
3. Jefferson, T., Rivetti, D., Rivetti, A., Rudin, M., Di Pietrantonj, C., Demicheli, V.: Efficacy and Effectiveness of Influenza Vaccines in Elderly People: A Systematic Review. Lancet 366 (2005) 1165-1174
4. McElhany, J.E.: Influenza: A Preventable Lethal Disease. J. Gerontol. A Biol. Sci. Med. Sci. 57 (2002) M627-M628
5. Meltzer, M. I., Cox, N. J., Fukuda, K.: The Economic Impact of Pandemic Influenza in the United States: Priorities for Intervention. Emerg. Infect. Dis. 5 (1999) 659-671
6. Muller, A. Association between Income Inequality and Mortality among US States: Considering Population at Risk. Am. J. Public Health 96 (2006) 590-591
7. Naumova, E.N., Castranovo, D., Cohen, S.A., Kosheleva, A., Naumov, Y.N., Gorski, J.: The Spatiotemporal Dynamics of Influenza Hospitalizations in the United States Elderly. In preparation
8. Nichol, K. L., Margolis, K. L., Wouremna, J., von Sternberg, T.: Effectiveness of Influenza Vaccine in the Elderly. Gerontology 42 (1996) 274-279

9. Piedra, P. A., Gaglani, M. J., Kozinetz, C. A., Herschler, G., Riggs, M., Griffith, M., Fewlass, C., Watts, M., Hessel, C., Cordova, J. ,Glezen, W. P.: Herd Immunity in Adults Against Influenza-related Illnesses with Use of the Trivalent-live Attenuated Influenza Vaccine (CAIV-T) in Children. Vaccine 23 (2005) 1540-1548

10. Reichert, T. A., Sugaya, N., Fedson, D. S., Glezen, W. P., Simonsen, L., Tashiro, M.: The Japanese Experience with Vaccination Schoolchildren against Influenza. N. Engl. J. Med. 344 (2001) 889-896

11. Simonsen, L., Reichert, T.A., Viboud, C., Blackwelder, W. C., Taylor, R. J., Miller, M. A.: Impact of Influenza Vaccination on Seasonal Mortality in the US Elderly Population. Arch. Intern. Med. 165 (2005) 265-272

12. Smith, N. M., Bresee, J.S., Shay, D. K., Uyeki, T. M., Cox, N. J., Strikas, R. A.: Prevention and Control of Influenza: Recommendations of the Advisory Committee on Immunization Practices (ACIP). MMWR Recomm. Rep. 55 (2006) 1-42

13. Thompson, W. W., Shay, D. K., Weintraub, E., Brammer, L., Cox, N., Anderson, L. J., Fukuda, K.: Mortality Associated with Influenza and Respiratory Syncytial Virus in the United States. JAMA 289 (2003) 179-86

14. Thompson, W.W., Shay, D. K., Weintraub, E., Brammer, L., Bridges, C.B., Cox, N.J., Fukuda, K.: Influenza-Associated Hospitalizations in the United States. JAMA. 292 (2004) 1333-1340

15. US Census Bureau: 2005 American Community Survey Gulf Coast Area Data Profiles: Louisiana Data Profiles.
    http://www.census.gov/acs/www/Products/Profiles/gulf_coast/tables/
    tab1_katrina04000US22v.htm. Accessed 3/12/07

16. Xu, K.T.: State-level Variations in Income-Related Inequality in Health and Health Achievement in the US. Soc. Sci. Med. 63 (2006) 457-464

Data Classification for Selection of Temporal Alerting Methods for Biosurveillance

Howard Burkom and Sean Murphy

National Security Technology Department
The Johns Hopkins University Applied Physics Laboratory,
Laurel, Maryland USA
{Howard.Burkom,Sean.Murphy}@jhuapl.edu

Abstract. This study presents and applies a methodology for selecting anomaly detection algorithms for biosurveillance time series data. The study employs both an authentic dataset and a simulated dataset which are freely available for replication of the results presented and for extended analysis. Using this approach, a public health monitor may choose algorithms that will be suited to the scale and behavior of the data of interest based on the calculation of simple discriminants from a limited sample. The tabular classification of typical time series behaviors using these discriminants is achieved using the ROC approach of detection theory, with realistic, stochastic, simulated signals injected into the data. The study catalogues the detection performance of 6 algorithms across data types and shows that for practical alert rates, sensitivity gains of 20% and higher may be achieved by appropriate algorithm selection.

Keywords: data classification, biosurveillance, anomaly detection, time series.

1 Introduction

This paper introduces a method for classifying time series for routine health surveillance according to the anomaly detection algorithm that is likely to yield the best detection performance among a set of candidate methods. The classification process involves repeated trials using stochastic simulated signals injected into background data, detection theoretic evaluation methods, and mining of the results for simple selection criteria using a number of possible discriminants computed from data samples. To exploit this guidance for an appropriate choice of method, however, the health monitor needs only to calculate the chosen discriminants and apply the criteria.

1.1 Scope of Study: Automated Health Surveillance

This paper is designed to assist the monitoring of the univariate time series that result from syndromic classification of clinical records and also from nonclinical data such as filtered counts of over-the-counter remedy sales. Timely electronic records that capture these data are increasingly being collected by health departments and other agencies in order to monitor these data for outbreaks of unspecified disease or of specified disease before the confirmation of identifying symptoms, a monitoring process often denoted as *syndromic surveillance*. Principal objectives of this

D. Zeng et al. (Eds.): BioSurveillance 2007, LNCS 4506, pp. 59–70, 2007.

surveillance are to complement and extend physician sentinel surveillance at false alarm rates manageable for the public health infrastructure and also to help characterize public health threats regarding their location, progression, and population-at-risk. Several research initiatives are underway for the efficient use of large, evolving collections of electronic data for routine monitoring. The first step is the appropriate filtering, classification, and cleaning of the record sets to form data structures such as time series that can be inspected regularly for anomalies. Anomaly detection algorithms are then applied to these time series to detect early effects of outbreaks at low false alarm rates. Elements of these algorithms such as forecast methods and control chart parameters and thresholds depend on the time series background and on the type of signal to be detected. The methodology of this paper derives simple criteria for choosing among candidate algorithms. This methodology may be applied to preprocessed series such as rates or proportions, weekly aggregates, intervals between cases, and others; the current study treats only daily count data. Intended beneficiaries are local public health monitors using their own data streams as well as large system developers managing many disparate data types.

1.2 Issues in the Choice of Alerting Algorithms

Numerous, recent papers have presented and evaluated algorithms for biosurveillance-related anomaly detection. However, authors of these papers are rarely able to share their datasets and can often publish only limited information describing them, so it may be difficult for a health monitor to determine whether a published method will have the desired sensitivity and specificity on the data at hand. Some research papers offer multiple algorithms, and it can be unclear how to choose among these or combine their results. Accentuating this problem, health monitors at 2005-2006 conferences and workshops related to automated health surveillance repeatedly expressed the need for rapidly modifiable case definitions and syndromic filters. Such on-demand classifications rule out methods with model parameters derived from years of historic data. Impromptu case definition changes may lead to changes in the scale and cyclic or seasonal series behavior of resulting time series. We demonstrate that mismatched algorithms and data can result in significant, systematic loss of sensitivity at practical false alarm rates. Therefore, the automated selection of suitable alerting methods is necessary.

2 Methods

A multiple-step approach was chosen to determine algorithm selection criteria using replicable statistical methods. This section presents this approach and describes the datasets, candidate algorithms, and Monte Carlo testing and evaluation procedure of this study.

2.1 Proposed Approach for Data-Adaptive Method Selection

We applied the following steps to determining algorithm selection criteria:

a. Select datasets: We chose authentic and simulated collections of time series representative of a range of scales and behaviors observed in implementations of the

Electronic Surveillance System for the Early Notification of Community based Epidemics (ESSENCE) [1]. Research authors typically evaluate algorithms using their available data and their choices of spatial, temporal, and syndromic aggregation (i.e. clinical case definition). The data and the corresponding filtering decisions appropriate for an epidemiologist health monitor may produce time series much different from those used in literature evaluations.

b. Select candidate algorithms: A limited set of methods, explained in Section 2.3, was chosen to represent the effects of data models and control charts.

For each time series in each chosen dataset:

c. Calculate candidate discriminants: For this study, the mean, median, variance, skewness, kurtosis, and autocorrelation coefficients of lags 1 and 7 were calculated for each data stream.

d. Form multiple copies with realistic injected signals: For each copy of the series, a signal start time was randomly chosen, and a plausible stochastic signal, explained in section 2.4, was added to the data to simulate the effect of a disease outbreak. Signal start dates were restricted beyond the warmup periods of all selected algorithms.

e. Exercise all candidate alerting algorithms on the multiple data streams.

f. Tabulate algorithm performance measures: These measures were selected and averaged values chosen from ROC curves chosen to indicate algorithm performance at practical false alarm rates.

Once performance measures for all algorithms and data stream were tabulated, the final steps were:

g. Correlate algorithm performance measures with candidate discriminants
h. Infer algorithm selection criteria

2.2 The Study Datasets

We used datasets of both authentic and simulated time series. Authentic time series are preferable for their realistic background behavior but have the drawback that false alarms cannot be truly identified. Simulated datasets are artificial but have a compensating advantage for algorithm evaluation--their statistical properties are known, and data from the time intervals without injected outbreaks represent noise, so there can be confidence that the computed alert rate using the original data is a false alert rate. Both datasets may be freely downloaded from [2].

Authentic dataset
The authentic dataset was originally approved for public distribution at a conference workshop given by the first author in 2005. The data consist of 33 time series of aggregate daily counts of military clinic visits covering the 1402 days from 28Feb1994 until 30Dec1997. These daily counts are from an unspecified metropolitan area, and neither author had access to the original data records or any identifying information therein. They were collected in the course of a multi-agency project to select syndrome groupings of ICD-9-coded data for bioterrorism surveillance. The final groupings chosen may be found in the document at [3], which contains details of the project. The dataset used in this study were formed at a late stage of the project, and the groupings used to collect its time series were

approximately those given in the document. The same reference explains that the groupings range from "general symptoms of the syndrome group" to "specific diagnoses that ... occur infrequently or have very few counts". Thus, the resulting time series range from rich temporal structure with obvious seasonal behavior to sparse counts with many zeros. Mean counts range from below 0.05 to over 300.

Simulated dataset

The second set of time series was a collection of 24 sets of 2000 counts of randomly generated Poisson data weighted according to day-of-week patterns observed in various ESSENCE data sources. For count data, the time series variance typically increases with the mean, and Poisson distributions are often used to represent them. Three sets of day-of-week factors, including a control, and 8 mean values were used to generate the 33 series. The 8 mean values were those in {0.5, 1, 3, 5, 10, 20, 50, 100}, and the day-of-week factors are shown in Table 1. These patterns are the most frequent source of bias in daily biosurveillance data.

Table 1. Mean weekly visit count patterns observed in ESSENCE data

Weekly Pattern	Sunday	Monday	Tuesday	Wednesday	Thursday	Friday	Saturday
Control	1.00	1.00	1.00	1.00	1.00	1.00	1.00
Civilian Office Visits	0.15	2.15	1.55	0.95	0.95	0.95	0.30
Military Office Visits	0.10	1.55	1.50	1.35	1.20	1.15	0.15

2.3   Candidate Algorithms

A small set of candidate detection methods was chosen to represent methods used in recent literature on biosurveillance. Two adaptive control charts and a temporal scan statistic were applied to both pure count data and forecast residuals. Forecasts were made using a modified Holt-Winters method for generalized exponential smoothing, discussed below.

Sliding Z-score

A thorough discussion of control charts, used to detect out-of-control conditions in industrial processes, may be found in [4]. These charts are traditionally applied to monitor measured parameters for prompt indications of trouble. In the changing counts of population care-seeking behavior monitored in biosurveillance, there is no process that can be controlled and corrected, and developers have modified control chart methods to accept evolving data streams that violate expected statistical criteria such as normality and independence. These modifications sacrifice the well-defined chart output behavior, and alerting thresholds are derived though a combination of statistical correction and empirical observation.

The Xbar chart, perhaps most widely used, monitors subgroup means of process parameters relative to an overall mean and subgroup standard deviation determined in an initial analysis phase. We use a sliding z-score in which the subgroup mean is replaced by an individual observation x_t such as a daily diagnosis count, and the reference mean μ_t and standard deviation σ_t are calculated from recent counts in a sliding baseline. The test statistic is thus:

$$z_t = (x_t - \mu_t) / \sigma_t \tag{1}$$

To apply this statistic to sparse as well as rich data stream, we imposed a minimum of 0.5 for σ_t to avoid erratic behavior of z_t for baselines with few counts. For example, this restriction can be used to avoid alerts resulting from an isolated case if the baseline has no case counts.

The baseline period should be long enough to avoid excess fluctuation while short enough to capture recent seasonal behavior. A buffer interval of a week or less is often inserted between the baseline period and the test day. This statistic is closely related to the C1 and C2 algorithms [5] that use a 7-day baseline and buffer periods of 0 and 2 days, respectively. In the current study, baseline intervals of 7, 28, and 56 days are used with buffers of 2 and 7 days. The ROC analysis described below may be used to determine alerting thresholds.

Adaptive CUSUM Chart

The cumulative summation, or CUSUM chart, has proven timelier than the X-bar chart at detecting small mean shifts and has also been applied to biosurveillance. Again we use a version that tests single observations normalized with reference to a sliding baseline mean μ_t and standard deviation σ_t. The traditional CUSUM chart maintains running sums of differences both above and below the mean to detect anomalies in either direction. For the biosurveillance context, we use an upper sum S_H to look only for excessive counts. Furthermore, small differences are ignored; only differences at least $2k$ standard deviations above μ_t are counted for a fixed multiple k. A common practice is to set k at 0.5 to detect a shift of one standard deviation. In terms of the statistic z_t, the test statistic for the upper sum is then:

$$S_{H,t} = \max(0, z_t - 0.5 + S_{H,t-1}). \tag{2}$$

The initial value $S_{H,0}$ may be set to an estimate of μ_t or it may be set to the first data value with negligible effect. In this study we set baseline and buffer intervals as for the sliding z-score method above.

Temporal scan statistic

For a third algorithm, we used a form of the temporal scan statistic of [6]. This method is related to a moving average control chart in that it tests the sum of observed counts $Y_t(w)$ on day t in a sliding test window of length w days. If the expected number of counts in the window on day t is $E_t(w)$, then the test statistic is:

$$G_t(w) = Y_t(w) \ln[Y_t(w) / E_t(w)] - [Y_t(w) - E_t(w)], \text{ if } Y_t(w) > E_t(w) \tag{3}$$

This G-surveillance statistic is based on a generalized likelihood ratio test designed to yield a fixed background alert rate in continual surveillance for a specified time period. We derived the count expectations in two ways: using the sliding baseline mean as above for an 8-week baseline preceding the test window, and using the forecast method described next.

Data forecasting method

The methods above use the counts directly without accounting for systematic behaviors such as seasonal patterns, day-of-week effects, and holidays. Some published alerting methods [7, 8, 9] account for these behaviors with regression models and autoregressive filters of varying complexity. These models make predictions of daily counts, and residuals are obtained by subtracting the predictions from observed counts. The residuals are then treated as test statistics or used as inputs to control charts. In this study we used a Holt-Winters forecasting technique to represent these methods. This technique generalizes the exponential smoothing of data values to the updating of linear trends and cyclic effects. In [10], this forecast method was applied to 16 syndromic time series, and the resulting goodness-of-fit measures compared well to those of regression models.

In Holt-Winters smoothing, auxiliary series L_t, T_t, and S_t for the level, trend, and cyclic multiplier, for a cyclic period M, are updated along with the forecast $\hat{y}_{t.}$ At time step t, the k-step-ahead forecast is:

$$\hat{y}_{t+k} = (L_t + k\,T_t)\,S_{t+k-M} \tag{4}$$

In the updating equations each auxiliary term is a convex combination of its previous value and a calculation using present behavior. Constant smoothing coefficients α, β, and γ determine the weighting of current values versus past history in these recursive equation :

$$L_t = \alpha\,(Y_t / S_{t-M}) + (1 - \alpha)\,(L_{t-1} + T_{t-1}) \tag{5}$$

$$S_t = \gamma\,(Y_t / L_t) + (1 - \gamma)\,S_{t-M} \tag{6}$$

$$T_t = \beta\,(L_t - L_{t-1}) + (1 - \beta)\,T_{t-1} \tag{7}$$

We used these equations to compute forecasts for the time series of the study. Following the implementation in [10], we set the cyclic multiplier M = 7 in order to model day-of-week effects, and we avoided training on outliers by not applying the updating equations (i.e. temporarily setting $\alpha = \beta = \gamma = 0$) when the absolute forecast error was greater than 50% of the prediction. The updating was also suspended to avoid zero divisors in (5) and (6).

Composite methods

The 3 forecast-based alerting algorithms in this study were composite methods derived by applying respectively the sliding zscore, CUSUM, and Gscan methods on residuals of Holt-Winters (HW) forecasts. This process is analogous to the application of Page's test in [8] to combine loglinear regression with a CUSUM chart. The motivation is to reduce control chart alerting bias by first removing predictable cycles and trends from the data. Thus, the 6 algorithms in the study were the sliding z-score, CUSUM, Gscan, and the composite HW-zscore, HW-CUSUM, and HW-Gscan.

2.4 Target Signal Types and Calculation

To compare the performance of the chosen algorithms, we formed multiple datasets for repeated trials by adding one stochastic signal to the background time series for each trial. For each time series, daily counts for these signals were randomly drawn from an idealized data epicurve, a model of the effect of an outbreak on the time series. For this model we adopted the Sartwell lognormal incubation period model [11]. This approach to simulated data effects assumes that the number of outbreak-attributable data counts on a given day is proportional to the number of newly symptomatic attributable cases on that day—see [12] for a fuller discussion. Given an outbreak start date, the resulting distribution depends on 2 parameters, the mean ζ and standard deviation σ of the logarithm of the incubation period length. Like actual disease outbreaks, the data effect may be gradual or explosive, depending on the values adopted for ζ and σ. We used two parameter sets in this study. An explosive outbreak with most cases affecting the data on the 1st or 2nd day was modeled with sets of runs with $\zeta = 1$ and $\sigma = 0.1$. For gradual outbreaks with a median incubation of 10-11 days, we used $\zeta = 2.4$ and $\sigma = 0.3$.

For each trial, we injected a fixed number of cases chosen to make detection reasonably challenging for the algorithms. For the explosive outbreak runs of this study, the number of injected counts was 2 or 4 times the standard deviation of the detrended data. For the gradual outbreak runs in which the cases were spread over 2-3 weeks, the number of injected counts was 8 or 10 times the detrended standard deviation. For sparse time series, the number of injected counts was at least 5.

With the signal shape and number of cases N set, we generated the set of incubation periods with a set of N random lognormal draws and rounded each to the nearest day, just as each actual incubation period could be seen as a random draw from distributions of dosage and susceptibility. The number of cases injected for each day was the number of draws rounded to that day. The resulting trial set was then the original time series with the injected counts added to the outbreak days. For each trial, the start day for injection was chosen with a random uniform draw from the days in the dataset, with the restrictions that the injections begin late enough for the maximum 56-day warmup period among the tested algorithms but early enough to fit within the data time series (at least 14 days before the series end for gradual signals).

2.5 Evaluation with Repeated Trials

To test the performance of each algorithm on each time series in the study datasets, we generated 600 trial sets as described above for each signal type and applied the algorithms to all of these trial sets. Algorithm outputs were collected for all days, both with and without injected outbreak cases. Because the start of each simulated outbreak is known, precise measurement of algorithm performance was possible.

The ROC curve provides a systematic approach to quantifying the tradeoff of sensitivity and specificity. Traditionally, the health industry uses the terms "sensitivity" and "specificity" to measure the accuracy of a diagnostic test. Here, "sensitivity" refers to the probability of a positive test result when an event of interest occurs. In conventional epidemiology, the event is the presence of disease; in this context, it is the presence of additional cases attributable to a simulated outbreak. "Specificity" refers to the probability of a negative result when the event of interest does not occur; this quantity is 1 minus the probability of a false alert.

For algorithm evaluation of this study, we allowed the alerting threshold to vary, and we computed for each threshold an empirical detection probability (PD), or sensitivity, and a false alert probability (PFA). The ROC curve is usually defined as the plot of PD versus PFA for all thresholds tested. In this study, the PD was the fraction of all trials for which the algorithm output exceeded the threshold on at least one day with a positive number of injected cases. The PFA was the fraction of days on which the output exceeded the threshold when we applied the algorithm to the original time series.

A single indicator commonly used for summarizing algorithm performance is the area under the ROC curve. This area may be interpreted as the average PD over the range of thresholds tested. For practical evaluation of algorithm performance, one may use the PD at a few selected alert rates or may take the average PD over some reasonable PFA interval. For this study we used the area under the ROC curve between alert rates of 1 per 112 days and 1 per 14 days.

3 Results

The limited data types included in this study indicated that easily computed data discriminants could serve as indications of detection performance to automate and perhaps periodically update the choice of alerting algorithm. Among the discriminants tested so far, the data mean (and median for nonsparse data) and autocorrelation coefficients have seemed most correlated to algorithm performance.

Table 2 presents the detection probabilities calculated for gradual signals added to the authentic syndromic series, with the total number of injected cases equal to 8 times each series standard deviation. The listed probabilities are averaged over those found for alert rates fixed between 1 per 2 weeks and 1 per 16 weeks. For brevity, the only attributes listed are the series mean and its lag-1 and lag-7 autocorrelation coefficients (shown as AC1 and AC7). These attributes were computed over the entire series. Eight columns of mean detection probabilities are shown.

Columns 5 through 8 give probabilities obtained by applying the CUSUM, Gscan with a 3-day window (GSCN3), Gscan with a 7-day window (GSCN7), and the sliding z-score (ZSCR) to the Holt-Winters forecast residuals.

The last 4 columns show probabilities obtained by applying the same 4 methods to the naïve counts.. Dark shading indicated a probability that is within 0.01 of the maximum for the data series of its row. The lighter shading indicates a probability lower than the maximum by at least 0.01 but less than 0.05.

The AC1 and AC7 values in the table show that autocorrelation is present in nearly all of the time series with mean daily counts above 1. For all time series with mean value above 5, subtraction of the Holt-Winters forecast improves sensitivity. The HW-CUSUM gives the best performance for nearly all of the sparse time series, while the HW-GSCN3 algorithm is best overall when the mean exceeds 10. As noted above, the use of unlabelled authentic series has the disadvantage that some background alerts may be true signals.

More studies are required to choose appropriate algorithms for series with low mean values and especially for sparse ones with a median value of 0.

Table 2. Average probabilities of detection of injected gradual signals for authentic data streams, with backgrouind data attributes

Data Attributes				HW-forecast methods				Pure Count Methods			
name	Mean	AC1	AC7	CUSUM	GSCN3	GSCN7	ZSCR	CUSUM	GSCN3	GSCN7	ZSCR
Neuro_3	0.00	-0.01	-0.01	1.000	0.000	0.000	1.000	1.000	0.000	1.000	1.000
Lesion_3	0.05	-0.03	-0.02	0.998	0.981	0.996	0.955	0.999	0.981	0.985	0.954
Rash_3	0.05	0.00	0.02	0.996	0.814	0.996	0.949	0.996	0.814	0.977	0.950
UnxplDth_1	0.07	0.09	0.00	0.984	0.812	0.935	0.913	0.983	0.775	0.875	0.914
Shk_Coma_1	0.07	0.14	0.13	0.985	0.441	0.370	0.963	0.983	0.494	0.650	0.954
UnxplDth_2	0.07	0.13	-0.01	0.930	0.092	0.153	0.857	0.930	0.082	0.200	0.846
Lymph_3	0.12	0.06	-0.01	0.968	0.714	0.842	0.897	0.958	0.662	0.527	0.880
Fever_3	0.21	0.06	0.11	0.867	0.500	0.759	0.831	0.864	0.429	0.553	0.813
Resp_3	0.22	0.14	0.06	0.835	0.568	0.586	0.835	0.835	0.475	0.515	0.827
Bot_Like_3	0.39	0.03	0.03	0.789	0.774	0.741	0.766	0.761	0.744	0.794	0.754
UnxplDth_3	0.43	0.09	0.06	0.763	0.667	0.821	0.755	0.755	0.510	0.716	0.748
Hemr_ill_1	0.60	0.16	0.17	0.821	0.717	0.800	0.787	0.804	0.702	0.701	0.780
Gi_3	0.60	0.05	0.03	0.682	0.614	0.640	0.658	0.665	0.586	0.599	0.646
Lesion_1	1.40	0.28	0.30	0.937	0.760	0.900	0.893	0.933	0.847	0.917	0.890
Neuro_1	2.02	0.36	0.30	0.833	0.656	0.574	0.803	0.821	0.650	0.604	0.793
Rash_1	3.00	0.17	0.11	0.740	0.844	0.738	0.799	0.686	0.791	0.809	0.711
Bot_Like_1	3.15	0.26	0.24	0.618	0.507	0.426	0.649	0.652	0.631	0.690	0.654
Shk_Coma_2	5.38	0.26	0.19	0.637	0.635	0.484	0.710	0.588	0.564	0.498	0.622
Lymph_1	6.78	0.24	0.21	0.765	0.945	0.842	0.781	0.706	0.697	0.641	0.745
UGI_2	7.19	0.34	0.29	0.772	0.678	0.440	0.755	0.712	0.810	0.778	0.762
Bot_Like_2	13.51	0.40	0.38	0.897	0.975	0.930	0.879	0.745	0.824	0.861	0.771
Hemr_ill_2	14.20	0.32	0.28	0.783	0.855	0.804	0.789	0.682	0.750	0.759	0.735
UGI_1	14.94	0.37	0.33	0.775	0.908	0.814	0.751	0.708	0.597	0.499	0.714
Lesion_2	22.43	0.27	0.23	0.738	0.858	0.795	0.808	0.578	0.727	0.727	0.743
Neuro_2	35.36	0.37	0.32	0.916	0.979	0.923	0.854	0.800	0.892	0.916	0.805
LGI_1	38.08	0.42	0.39	0.821	0.876	0.774	0.763	0.699	0.648	0.505	0.703
Gi_1	53.02	0.42	0.40	0.845	0.904	0.822	0.762	0.689	0.651	0.520	0.704
LGI_2	54.92	0.36	0.33	0.895	0.977	0.949	0.873	0.825	0.897	0.929	0.865
Gi_2	62.11	0.36	0.33	0.915	0.985	0.963	0.889	0.840	0.912	0.927	0.888
Rash_2	77.91	0.31	0.27	0.896	0.978	0.947	0.836	0.809	0.905	0.927	0.840
Fever_1	78.98	0.51	0.45	0.770	0.691	0.616	0.656	0.662	0.473	0.282	0.664
Resp_2	161.99	0.35	0.32	0.893	0.969	0.914	0.840	0.834	0.805	0.834	0.850
Resp_1	334.61	0.52	0.50	0.875	0.932	0.837	0.834	0.799	0.458	0.442	0.825

An AC7 value above 0.2 indicates a day-of-week effect that can be included in a forecast method to improve algorithm performance. This performance correlation was seen even for subsyndrome values whose mean value is below 5 and for which day-of-week effects are not visually obvious in the time series.

For explosive outbreaks, the best performing methods were the z-score applied to the counts and to the HW residuals, seen in both the authentic and the simulated datasets.

Table 3 gives a portion of the table for the simulated data, including results for series with weekly patterns in Table 1. Without the weekly pattern, the HW forecasting with a positive cyclic smoothing coefficient is analogous to a regression model with a predictor variable that is uncorrelated with the outcome. Detection probabilities are generally low, and the methods based on pure counts outperform those based on residuals, though the PD penalty for misfitting the forecast never exceeds 0.08. For the runs on the two weekly patterns, however, the methods based

Table 3. Average probabilities of detection of injected rapid-onset signals for simulated data streams, with background data attributes

Data Attributes			HW-forecast methods				Pure Count Methods			
Mean	AC1	AC7	CUSUM	GSCN3	GSCN7	ZSCR	CUSUM	GSCN3	GSCN7	ZSCR
No Weekly Pattern										
0.5	-0.01	0.03	0.908	0.898	0.598	0.962	0.883	0.964	0.698	0.959
1.0	-0.02	0.02	0.674	0.720	0.354	0.897	0.776	0.855	0.464	0.906
3	-0.04	-0.05	0.433	0.499	0.255	0.569	0.509	0.552	0.283	0.645
5	-0.01	-0.01	0.298	0.340	0.236	0.424	0.372	0.421	0.251	0.475
10	-0.01	0.00	0.339	0.374	0.282	0.377	0.308	0.407	0.223	0.397
20	0.01	0.02	0.335	0.326	0.187	0.422	0.306	0.367	0.210	0.424
50	0.01	0.00	0.249	0.324	0.244	0.361	0.289	0.318	0.189	0.415
100	0.01	0.01	0.275	0.311	0.241	0.378	0.279	0.383	0.222	0.404
Weekly Pattern #1										
0.5	-0.02	0.16	0.819	0.798	0.566	0.849	0.796	0.802	0.556	0.833
1.0	0.01	0.17	0.630	0.429	0.151	0.720	0.467	0.516	0.375	0.622
3	0.03	0.43	0.421	0.533	0.288	0.556	0.248	0.293	0.298	0.295
5	0.05	0.57	0.451	0.631	0.314	0.696	0.348	0.395	0.392	0.315
10	0.05	0.76	0.648	0.810	0.437	0.833	0.352	0.420	0.477	0.314
20	0.08	0.85	0.792	0.863	0.566	0.875	0.441	0.461	0.567	0.328
50	0.09	0.93	0.942	0.975	0.816	0.978	0.538	0.499	0.883	0.322
100	0.09	0.96	0.994	1.000	0.990	0.998	0.596	0.554	0.986	0.344
Weekly Pattern #2										
0.5	0.02	0.07	0.922	0.939	0.719	0.938	0.903	0.913	0.742	0.926
1.0	0.09	0.18	0.582	0.461	0.254	0.772	0.586	0.565	0.411	0.716
3	0.11	0.41	0.538	0.573	0.305	0.687	0.381	0.421	0.329	0.478
5	0.14	0.55	0.435	0.566	0.205	0.611	0.331	0.384	0.292	0.408
10	0.20	0.72	0.730	0.761	0.474	0.793	0.503	0.474	0.446	0.546
20	0.26	0.83	0.820	0.881	0.586	0.907	0.552	0.535	0.574	0.642
50	0.28	0.92	0.934	0.972	0.779	0.994	0.650	0.564	0.882	0.724
100	0.29	0.96	0.989	0.999	0.927	0.999	0.714	0.572	0.967	0.732

on residuals have substantially higher detection probabilities than those based on pure counts, with PD increases of over 0.2 even for mean values as low as 3. However, when the data become sparse and AC7 drops substantially below 0.2, the advantage of the forecasting becomes negligible.

For the authentic data, results were generally similar for rapid-onset outbreaks. The residual-based z-score PD values were uniformly above those of the count-based z-score, and the advantage exceeded 0.2 for 12 of the 16 series with means above 5.

4 Discussion

The study implemented here on a selection of time series from a single data source and a from a set of random draws from known distributions suggests that readily computed measures from data samples may guide the selection of alerting algorithms for biosurveillance. The indicated algorithm choice depends on the signal type of interest. For the detection of explosive signals, a sliding z-score method patterned after the Xbar chart outperformed the other candidate methods. For mean values above 5 and AC7 values above 0.2, the z-score applied to forecast residuals was superior, while for sparser data, a z-score applied to the pure counts was indicated. For gradual signals, the results suggest that a temporal scan statistic using a 3-day window of forecast residuals is a good choice for $AC7 > 0.2$ and mean > 5, and that a CUSUM applied to data counts is indicated for sparser data.

These results must be kept in context. Alternative forecast methods and control chart ideas may improve the sensitivities and more informative data measures may be found. Holt-Winters forecasting has the advantages that it requires little data history, is more versatile than regression methods at adjusting to local linear trends [10, 13] and year-to-year seasonal variation, and may be implemented on a spreadsheet. In a detailed comparison of this approach to autoregressive methods, [13] points out that Box-Jenkins linear filter forecasts should be superior if the time series of interest can be analyzed for optimal model parameters. However, given the demands for adjustable case definitions that determine the time series of interest, the challenge is to model linear filters promptly with limited data history. In the modeling work of [14], two autoregressive moving average models were robust predictors, and forecasts of these models should be compared to HW forecasts for a range of data types.

For using calculated data measures for algorithm selection, a practical issue is the size of the data sample required to compute them. Preliminary work suggests that a month of data counts may be sufficient for rich, structured time series, but longer samples or expert knowledge are probably needed for series with low counts. As data history accumulates and as time series behavior evolves, periodic reclassification of monitored data streams may be necessary for ongoing, robust detection performance.

The monitoring of sparse data streams with mean daily counts below 1 is important for surveillance efforts in which a large set of rare subsyndrome groups or counts from many small regions such as zip codes must be monitored. Suitable algorithm selection can help retain sensitivity while controlling the multiple-testing problem in such situations. The methods of this study need to be modified for sparse time series; larger repeated trial sets and a modified signal simulation procedure should be considered.

5 Conclusions

Algorithm/data-type mismatches can degrade the effectiveness of a detection system; large differences in detection probability were observed among the candidate alerting methods. This preliminary study presents evidence that the application of simple discriminants can help ensure selection of alerting methods well suited to the data. This process can be improved, applied to additional algorithms, and extended for other data types in an evolving, shared research effort.

For future studies of this nature, the analysis of multiple sets of Monte Carlo runs to evaluate various data measures for data classification performance presented a challenging, multidimensional data mining problem, but one whose solution can provide versatile, easily automated guidance.

Acknowledgments. The authors acknowledge the help of Michael Thompson of the Johns Hopkins APL for providing the day-of-week calculations in Table 1.

References

1. Lombardo JS, Burkom HS, Elbert YA, Magruder SF, Lewis SH et al (2003). "A Systems Overview of the Electronic Surveillance System for the Early Notification of Community-Based Epidemics", Journal of Urban Health, Proceedings of the 2002 National Syndromic Surveillance Conference, Vol. 80, No. 2, Supplement 1, April 2003, New York, NY, p.i32-i42.
2. Data download: see online research site for the International Society for Disease Surveillance: https://wiki.cirg.washington.edu/pub/bin/view/Isds/ResearchTopics
3. See groupings at: www.bt.cdc.gov/surveillance/syndromedef/word/syndromedefinitions.doc
4. Ryan TP. Statistical Methods for Quality Improvement. New York: John Wiley & Sons: New York, 1989
5. Hutwagner L, Browne T, Seeman GM, Fleischauer AT. Comparing aberration detection methods with simulated data. Emerg Infect Dis [serial on the Internet]. 2005 Feb [date cited]. Available from http://www.cdc.gov/ncidod/EID/vol11no02/04-0587.htm
6. Wallenstein S, Naus Jhttp Scan Statistics for Temporal Surveillance for Biologic Terrorism. www.cdc.gov/MMWR/preview/mmwrhtml/su5301a17.htm
7. Kleinman K, Lazarus R, Platt R. A generalized linear mixed models approach for detecting incident clusters of disease in small areas, with an application to biological terrorism. Am J Epidemiol 2004;159:217-24
8. Reis BY, Mandl KD, Time series modeling for syndromic surveillance (2003). BMC Medical Informatics and Decision Making 2003, 3:2.
9. Brillman JC, Burr T, Forslund D, Joyce E, Picard R and Umland E. Modeling emergency department visit patterns for infectious disease complaints: results and application to disease surveillance, BMC Medical Informatics and Decision Making 2005, 5:4, pp 1-14
10. Burkom HS, Murphy SP, Shmueli G, Automated Time Series Forecasting for Biosurveillance (2007), Statistics in Medicine (accepted for 2007 publication)
11. Sartwell, PE. The distribution of incubation periods of infectious disease. Am J Hyg 1950; 51:310-318
12. Burkom, H., Hutwagner, L., Rodriguez, R. (2005), "Using Point-Source Epidemic Curves to Evaluate Alerting Algorithms for Biosurveillance," 2004 Proceedings of the Am. Statl Assoc., Statistics in Government Section
13. Chatfield C. The Holt-Winters Forecasting Procedure. App Stats 1978; 27: 264-279.
14. Reis BY, Pagano M, and Mandl KD. Using temporal context to improve biosurveillance. PNAS 100: 1961-1965; published online before print as 10.1073/pnas.0335026100

High Performance Computing for Disease Surveillance

David Bauer[1], Brandon W. Higgs[1], and Mojdeh Mohtashemi[1,2]

[1] The MITRE Corporation,
7515 Colshire Drive, McLean VA 22102
{dwbauer,bhiggs,mojdeh}@mitre.org
http://www.mitre.org/
[2] MIT CS and AI Laboratory,
32 Vassar Street, Cambridge MA 02139
mojdeh@mit.edu
http://www.mit.edu/

Abstract. The global health, threatened by emerging infectious diseases, pandemic influenza, and biological warfare, is becoming increasingly dependent on the rapid acquisition, processing, integration and interpretation of massive amounts of data. In response to these pressing needs, new information infrastructures are needed to support active, real time surveillance. Space-time detection techniques may have a high computational cost in both the time and space domains. High performance computing platforms may be the best approach for efficiently computing these techniques. Our work focuses on efficient parallelization of these computations on a Linux Beowolf cluster in order to attempt to meet these real time needs.

Keywords: HPC, High Performance Computing, Parallel Computing, Disease Surveillance, Beowolf cluster.

1 Introduction

Timely detection of infectious disease outbreaks is critical to real time surveillance. Space-time detection techniques may require computationally intense search in both the time and space domains [5,6]. The real time surveillance constraints dictate highly responsive models that may be best achievable utilizing high performance computing platforms. We introduce here a technique that performs efficiently in parallel on a Linux Beowolf Cluster.

In the special case of parallelization, Amdahl's law [1] states that if F is the portion of an application that is sequential, then the maximum speedup S that can be achieved using N processors is given by Equation 1.

$$S_{max} = \frac{1}{F + \frac{(1-F)}{N}} \ . \tag{1}$$

As $N \rightarrow \infty$, S_{max} approaches $1/F$, and thus the amount of improvement is limited by the sequential portion of the application. Conversely, as $F \rightarrow 0$,

D. Zeng et al. (Eds.): BioSurveillance 2007, LNCS 4506, pp. 71–78, 2007.

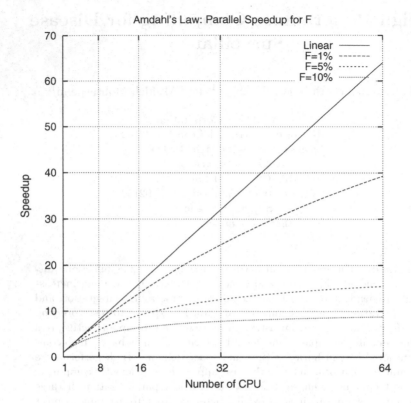

Fig. 1. Amdahl's Law illustrates effective speedup

S_{max} goes to N, that is the ideal parallelization, where each additional processor contributes fully to the performance of the application. In practice, this is a difficult goal to achieve. Efficient parallelization requires F be small as N increases. Figure 1 illustrates that, for $N = 64$ processors and $F = 10\%$ sequential, that the expected speedup approaches, but cannot exceed a 10-fold improvement in the running time. Finally, we define the *efficiency* of the parallelization by the Equation 2:

$$E_N = \frac{S_{max}}{N} .$$

(2)

The traditional approaches towards parallelizing an application fall into two categories: *(i) data and (ii) algorithm decomposition.* Because of the high degree of data dependency within the scan statistic algorithm, we focus on decomposing the application in the data domain. We modify the scan statistic data domain by computing individual addresses, thereby making the scope of the calculation significantly larger, based on the $O(N^2)$ algorithm complexity. While the overall running time of the application is larger on a single CPU, we will show that it can be computed in less time utilizing multiple processors executing in parallel by minimizing F.

The scan statistic was first introduced by Naus for cluster detection in 1965 [7] and later implemented and refined in other work [4,5,6,9,10], primarily for aberrant event detection such as those provided for surveillance purposes. The idea relies on a scanning window that utilizes multiple overlapping cylinders, each composed of both a space and time block, where time blocks are continuous windows (i.e., non-intermittent) and space blocks are geographic encompassing circles of varying radii, such as zip codes or census tracts.

In this paper, we focus our modeling efforts on clustering individual addresses as opposed to agglomerative regions (i.e. zip codes, census tracts, etc.) while using a derivation of the scan statistic [7] for disease surveillance in a metropolitan population. The use of individual addresses with the modified scan statistic is found to be a more sensitive measure for detection than census tract centroids. The former method also provides smaller time windows for significantly detected signals, which can speed the response time of an outbreak.

2   An Application: Space-Time Permutation Scan Statistic

Variations to both the scan statistic introduced by [6] and the method for fast detection of spatial over-densities, provided by [8], is implemented here as a suitable method for early detection of outbreaks in the metropolitan population, particularly for those time/region-specific increases in case frequency that are too subtle to detect with temporal data alone. Similar to the overlapping windows in the method proposed by [6], the scanning window utilizes multiple overlapping cylinders, each composed of both a space and time block, where time blocks are continuous windows (i.e. not intermittent) and space blocks are geographic-encompassing regions of varying size. However, instead of circles of multiple radii, a square grid approach, similar to that provided by [8] is implemented here.

Briefly explained below (see [6] for complete algorithm details), for each grid element, the expected number of cases, conditioned on the observed marginals is denoted by μ where μ is defined as the summation of expected number of cases in a grid element, given by Equation 3,

$$\mu = \sum_{(s,t)\in A} \mu_{st} \ . \tag{3}$$

where s is the spatial cluster (e.g., zip codes, census tracts, individual addresses) and t is the time span used (e.g., days, weeks, months, etc.) and

$$\mu_{st} = \frac{1}{N} \left(\sum_s n_{st}\right) \left(\sum_t n_{st}\right) \ . \tag{4}$$

where N is the total number of cases and n_{st} is the number of cases in either the space or time window (according to the summation term). The observed number of cases for the same grid element is denoted by n. Then the Poisson generalized likelihood ratio (GLR), which is used as a measure for a potential outbreak in the current grid element, is given by Equation 5 [4]:

$$\left(\frac{n}{\mu}\right)^n \left(\frac{N-n}{N-\mu}\right)^{(N-n)}. \tag{5}$$

Since the observed counts are in the numerator of the ratio, large values of the GLR signify a potential outbreak. To assign a degree of significance to the GLR value for each grid element, Monte Carlo hypothesis testing [2] is conducted, where the observed cases are randomly shuffled proportional to the population over time and space and the GLR value is calculated for each grid element. This process of randomly shuffling is conducted over 999 trials and the random GLR values are ranked. A p-value for the original GLR is then assigned by where in the ranking of random GLR values it occurs.

3   Performance Study

3.1   Computing Testbed and Experiment Setup

The Hive cluster at MITRE is a Red Hat Enterprise Linux 9.0 cluster consisting of 16 machines or compute nodes, for a total of 64 processors. The nodes are inter-connected via a dedicated gigabit ethernet switch. Each node's hardware configuration consists of a dual-processor, dual-core AMD Opteron 275 server and 8GBs of main memory. The AMD Opteron 200-series chip enables 64-bit computing, and provides up to 24GB/s peak bandwidth per processor using HyperTransport technology. The DDR DRAM memory controller is 128-bits wide and provides up to 6.4GB/s of bandwidth per processor. Our RAM configuration consisted of 4 2GB sticks of 400MHz DDR ECC RAM in 8 banks.

3.2   Model Scenario

The scanning window in our scan statistic model utilizes multiple overlapping grid elements, where the space blocks may be either individual addresses, or census tracts. We present results for both scenarios as a comparison of the effective amount of parallelism available in the data domain.

The San Francisco Department of Public Health (SFDPH), Tuberculosis Program provided the data. The geospatial information in the data consist of precise locations of 392 homeless individuals infected with tuberculosis (TB). The primary residences of these individuals were used to identify their geographical coordinates (latitudes and longitudes) using ArcGIS v9.0 (ESRI). The census tract information for identifying the tracts in which the homeless individuals reside, were obtained from generalized extracts from the Census Bureau's TIGER geographic database provided by the US Census Bureau (http://www.census.gov/geo/www/cob/index.html). The total number of unique census tracts for our metropolitan area is taken to be 76.

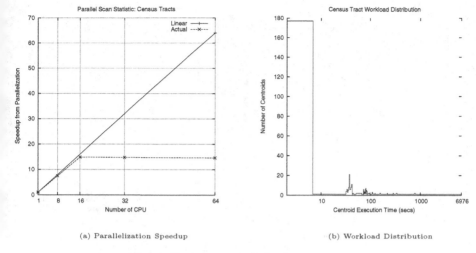

(a) Parallelization Speedup   (b) Workload Distribution

Fig. 2. Performance results for census tracts

For our space window, we restricted the geographic squares to sizes ranging from 0.02 km to 1 km in size, where for each separate sized-square, a neighboring square was allowed to overlap at half of the width on each side. For different sized squares that had a perfect intersect of the same cases, the smallest square was retained. The total number of space squares sampled for the census tract centroids was 441, while the total for individual residences was 4,234. For our time window case counts were used with time windows of 4 to 72 weeks spanning a period of ten years.

The scan statistic model is thus parallelized by mapping the individual spaces onto CPUs for processing. We utilize the Unified Search Framework (USF) developed at RPI for spawning the jobs onto the cluster compute nodes, and collecting the overall runtime of the model [11]. Our expectation is that the spatial blocks will exhibit a uniformly distributed workload, and that will translate into a highly efficient parallel model, although we cannot know the distribution until runtime.

3.3   Census Tract Performance

For comparison purposes, we first perform a parallelization of the scan statistic using census tracts. Our model consists of 441 centroids, and we computed mappings over 1, 8, 16, 32 and 64 processors. The purpose is to illustrate the effective scalability over a varying number of processors.

Figure 2a shows that using 16 processors achieves an efficiency of 93%, reaching an almost 15-fold improvement over sequential. When we utilize 32 or 64 CPUs, the efficiency drops to under 45%, indicating that there is a large sequential component to the data set once the runtime improves to around 7,000 seconds.

The issue here is that the workloads are not uniform in terms of computational workload, with some spatial squares requiring far more effort. The workload distribution is described by the histogram displayed in Figure 2b, and the longest running job last 6,976 seconds. In the 8 and 16 CPU cases, the running times

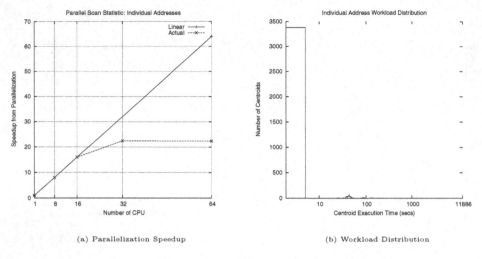

(a) Parallelization Speedup   (b) Workload Distribution

Fig. 3. Performance results for individual addresses

of all other centroids consumed at least this much time, and so the efficiency of the computation is high. The 32 and 64 CPU cases are unable to improve on the running time of the longest centroid, and so the efficiency drops as more CPU are added. To keep the efficiency high would require either increasing the total workload in the system, or parallelization within the scan statistic algorithm for the longest running centroids.

Although the efficiency of the parallelization limits us to making effective use of no more than 16 CPUs, we do report a significant improvement on the overall runtime of the algorithm for this data decomposition. The runtime dropped from over 28 hours to just under 2 hours.

3.4   Individual Address Performance

Decomposing the data domain into individual addresses allows for a higher degree of parallelism within the application. As with census tracts, the efficiency of the parallelization is high up to 16 CPUs. Figure 3a shows an efficiency of 68.75% for 32 CPU, an improvement over census tracts, and a 22-fold improvement in the runtime.

It is clear from the histogram in Figure 3b that the workloads based on individual addresses are more uniformly distributed, allowing for more efficient use of the cluster resources. However, the overall running time of the parallel application continues to be dominated by the most computationally complex centroid. Because the runtime of the longest census tract is smaller than the runtime of the longest individual address (6,977 seconds vs. 11,886), the individual address decomposition is not able to complete in less time on the cluster. The overall runtime for either are relatively close; execution times for all experiments are shown in Table 1.

Table 1. Parallel Execution Performance of the Scan Statistic Model

Data Set	# CPU	Runtime in Seconds	Speedup	Efficiency (%)
Census	1	103,129	-	-
Census	8	13,672	7.54	94.25
Census	16	6,940	14.86	92.875
Census	32	6,977	14.78	46.18
Census	64	7,055	14.61	22.82
Individual	1	267,029	-	-
Individual	8	33,305	8.0	100
Individual	16	14,401	16.0	100
Individual	32	11,886	22.4	70.18
Individual	64	11,918	22.4	35.0

The sequential running time for individual addresses is nearly 3 times larger than for the census tract data, from 28 to 74 hours. The parallelization improvement yields a decrease in the runtime of the model, to 3.3 hours.

4 Footnote on Parallelism

In [3], researchers at Sandia illustrate that the size of the model should be scaled with the number of processors utilized. By fixing the size of the model, it would appear that Amdahl's law would dictate only *embarrassingly parallel* (F very close to 0) applications would see a speedup using more than 100 CPUs. However, whenever additional computing capacity is available, it is more common that the size of the problem grows to consume that capacity. It may be more meaningful to propose a fixed *runtime*, rather than a fixed *problem size*.

In fact, this situation is encountered here. The size of our data set has not been scaled with the number of processors utilized. If we had done so, then the efficiency of the parallelization would have been much closer to 100% for the 32 and 64-CPU cases. This stems from the fact that the sequential portions of the application (time for start-up, serial bottlenecks, etc) do not increase as the problem size increases.

5 Conclusions and Future Work

We have proposed a highly efficient parallel computation technique for the real time surveillance of infectious disease outbreak detection. We have shown that high performance computing platforms, such as a Linux Beowolf Cluster, can come closer to meeting the needs of real time surveillance constraints.

In this paper we have reported results on the parallelization of a scan statistic model that has been modified to compute individual addresses as well as agglomerate regions, such as census tracts. We have shown that while the sequential execution time of the model is significantly larger, we can equalize the magnitude of the running time through parallelization of the model.

In the future we plan to investigate methods of parallelizing the scan statistic algorithm, to further increase the efficiency of the model on multiple processors. One expected outcome of this effort would be to balance the workload within a single centroid over multiple processors, and thus balance the non-uniform workload across the cluster computer. Our intent would be to improve the efficiency of the parallel scan statistic model.

References

1. Amdahl, G.: Validity of the Single Processor Approach to Achieving Large-Scale Computing Capabilities. In AFIPS Conference Proceedings (1967) 30:483-485
2. Dwass M.: Modified randomization tests for non-parametric hypotheses. In The Annals of Mathematical Statistics (1957) 29:181187
3. Gustafson, J.: Reevaluating Amdahl's Law. In Communications of the ACM (1988) 31(5):532-533
4. Kleinman KP, Abrams AM, Kulldorff M, Platt R: A model-adjusted space-time scan statistic with application to syndromic surveillance. In Epidemiology and Infectection (2005) 000:1-11
5. Kulldorff M.: A spatial scan statistic. Communications in Statistics: Theory and Methods (1997) 26:1481-1496
6. Kulldorff M, Heffernan R, Hartmann J, Assuncao R, Mostashari F: A space-time permutation scan statistic for disease outbreak detection. In Public Library of Science (2005) 2(3)
7. Naus J.: The distribution of the size of maximum cluster of points on the line. In Journal of the American Statistical Association (1965) 60:532-538
8. Neill DB, Moore AM.: A fast multi-resolution method for detection of significant spatial disease clusters. In Advances in Neural Information Processing Systems (2003) 16
9. Wallenstein S.: A test for detection of clustering over time. In American Journal of Epidemiology (1980) 111:367-372
10. Weinstock MA.: A generalized scan statistic test for the detection of clusters. In International Journal of Epidemiology (1982) 10:289-293
11. Ye T., Kalyanaraman S.: A Unified Search Framework for Large-scale Black-box Optimization. Rensselaer Polytechnic Institute, ECSE Department, Networks Lab (2003)

Towards Real Time Epidemiology: Data Assimilation, Modeling and Anomaly Detection of Health Surveillance Data Streams

Luís M.A. Bettencourt[1], Ruy M. Ribeiro[1], Gerardo Chowell[1], Timothy Lant[2], and Carlos Castillo-Chavez[3]

[1] Theoretical Division and Center for Non-Linear Studies,
Los Alamos National Laboratory,
MS B284, Los Alamos NM 87545, USA
[2] Decision Theater, Arizona State University,
PO Box 878409, Tempe AZ, 85287-8409, USA
[3] Department of Mathematics and Statistics, Arizona State University,
PO Box 871804, Tempe AZ, 85287-1804, USA

Abstract. An integrated quantitative approach to data assimilation, prediction and anomaly detection over real-time public health surveillance data streams is introduced. The importance of creating dynamical probabilistic models of disease dynamics capable of predicting future new cases from past and present disease incidence data is emphasized. Methods for real-time data assimilation, which rely on probabilistic formulations and on Bayes' theorem to translate between probability densities for new cases and for model parameters are developed. This formulation creates future outlook with quantified uncertainty, and leads to natural anomaly detection schemes that quantify and detect disease evolution or population structure changes. Finally, the implementation of these methods and accompanying intervention tools in real time public health situations is realized through their embedding in state of the art information technology and interactive visualization environments.

Keywords: real time epidemiology, data assimilation, Bayesian inference, anomaly detection, interactive visualization, surveillance.

1 Introduction and Motivation

Surveillance systems collect, analyze and report data continuously. Recent progress in information science and technology is increasingly enabling the collection of public health data worldwide in near real-time. In the United States, partly motivated by bio-security concerns, real-time bio-surveillance systems, that follow direct and indirect indicators of (epidemic) public health outbreaks have been developed nationwide, at the city level e.g. for New York City or Los Angeles, and at the state level such as in Michigan, Utah or Ohio. The monitoring of indirect data streams, pertaining e.g. to work or school absenteeism, emergency room and physician office calls, sales of certain over-the-counter medicines, known as syndromic surveillance [1,2,3,4] is also underway.

D. Zeng et al. (Eds.): BioSurveillance 2007, LNCS 4506, pp. 79–90, 2007.
© Springer-Verlag Berlin Heidelberg 2007

The use of a large number of real time data streams to infer the status and dynamics of the public health of a population presents enormous opportunities as well as significant scientific and technological challenges. A recent committee report from the Institute of Medicine of the National Academies concluded: "... that steps [must] be taken now to adapt or develop decision-aid models that can be readily linked to surveillance data to provide real-time feedback during an epidemic" [17]. Among the "Grand Challenges in Global Health" (http://www.gcgh.org/), articulated by a consortium of international organizations such as the Bill and Melinda Gates Foundation, The Wellcome Trust and the Canadian Institutes of Health Research, one finds the development of "... technologies that permit quantitative assessment of population health status". Naturally, the early identification and detection of emerging pathogens, pandemics or bioterrorist attacks call for the development and deployment of real-time syndromic surveillance systems.

Real time public health surveillance has as its primary mission pre-emptying the successful invasion or establishment of emerging infectious diseases. Its goals include the prevention or mitigation of epidemic growth, if possible by setting quantitative targets for intervention as events unfold and by providing a running time expectation for logistic allocations, including medicines, hospital beds and vaccines, if available. The success of real time public health surveillance depends on our ability to develop not only effective detection systems but also ways of evaluating the uncertainties associated with outbreaks, methods and models.

To a great degree, the reliability of a surveillance system hinges on its ability to make the extrapolations needed to predict the likely course of a public health emergency given the incoming streaming information. Hence, the need to develop statistical models of data assimilation that effectively estimate the changing parameters that characterize models of disease *dynamics*. In this paper a general procedure for data assimilation from real time data streams that feeds into epidemic models of communicable disease spread is developed. This procedure allows i) the estimation of probability densities for epidemiological parameters (such as disease transmissibility) and ii) the *prediction* of future observables with a quantified degree of confidence. This dynamical estimation/extrapolation environment lends itself naturally to a statistical anomaly detection scheme capable of identifying shifts in the public health status of a population, either via pathogen evolution, or as the result of changes in the population structure.

2   Data Assimilation and Probabilistic Prediction

Real time estimation of epidemiological parameters is conditioned by the types of data available as well as the target quantities for estimation (Figure 1). It is essential to produce running time statistics that incorporate past and present measurements as well as ways to assimilate these data to generate probabilistic predictions of new cases with quantified uncertainty because often the objective of real time surveillance is to detect change. This approach also allows the use of statistical estimation to partially bypass issues of non-stationarity in the data,

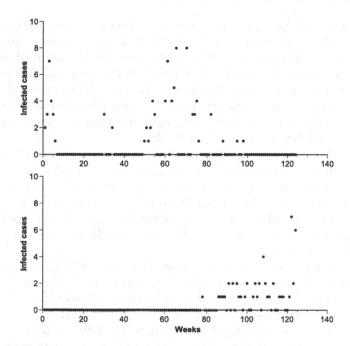

Fig. 1. Number of cases of H5N1 human influenza. Cases confirmed by the World Health Organization for Vietnam (top) and Indonesia (bottom), from January 2004 to June 2006. Although these are currently the two countries with most recorded cases, the incidence times series is very stochastic and intermittent (note the y-scale).

which typically hinder purely statistical (stationary) approaches to syndromic surveillance. Technically, this can be achieved by adopting a probabilistic approach to the *prediction* of new cases (or their proxy variables) at a given time t, $\Delta C(t)$, given past incidence time series and a model for the disease dynamics written in terms of a set of parameters Γ. The set Γ includes familiar parameters such as measures of transmissibility (e.g. the reproduction number R) and the duration of the infectious and incubation periods.

Here we focus on the estimation of a key quantity in epidemiology known as the reproduction number or ratio which quantifies the transmissibility of infectious diseases under identifiable initial conditions. In the context of novel infectious pathogens (e.g., the 1918-19 influenza pandemic, or the 2002-03 SARS epidemic) to which most of the population is completely susceptible, the *basic* reproduction number (denoted by R_0) quantifies the number of secondary cases generated by a primary case during its period of infectiousness within a *purely* susceptible population. In more practical situations, the population's actual susceptibility is affected by recurrent exposures to the infectious agent (e.g., dengue fever, influenza), vaccination campaigns (e.g., influenza, measles, mumps) and by "depletion" of susceptibles during the course of an outbreak. Then, the reproduction number denoted by R accounts for the residual immunity in the population. The relationship between R_0 and R *may* be modeled by $R = (1 - p)R_0$ where p

denotes the proportion of the population that is effectively protected to infection, assuming a well-mixed population [5,7,18,19,20]. In the advent of an epidemic, the timely estimation of the reproduction number with the minimal amount of available data crucial as it would allow public health authorities to determine the types and intensity of interventions necessary to achieve *fast* epidemic control. The level of critical vaccination coverage needed to diminish or eliminate various infectious diseases have also been determined from the magnitude of the estimated reproduction number [5,7,18,19,20].

A number of approaches have been used to estimate the reproduction number of infectious diseases including trajectory matching via least square fitting of epidemic models to time series epidemic data (e.g., [9,10,11]), methods that rely on the final epidemic size relation with the reproduction number (e.g., [13,15,6]), methods that make use of serological survey data (e.g., [14,12]), and recent probabilistic approaches that estimate the *effective* reproduction number over time (e.g., [8,21,22,23,24,16]).

The dynamical probabilistic approach described here [8,21,22,23,24,16] allows for the estimation of parameter density distributions via Bayes' theorem. Furthermore, as byproducts, the method leads to the automatic uncertainty quantification of predictions and a scheme for anomaly detection. The probabilistic prediction for new cases ($\Delta C(t)$) is written in terms of the probability distribution of given previous incidences ($\Delta C(t - \tau)$) and dynamical parameters, which we denote by

$$P\left[\Delta C(t) \leftarrow \Delta C(t - \tau) | \Gamma\right] = P\left[\Delta C(t) | \Delta C(t - \tau), \Gamma\right]. \tag{1}$$

This distribution encapsulates the dynamical prescriptions from different dynamical models of (epidemiological) disease dynamics. Below we show how this formulation can be specified in practice for different models, how it implies the full distribution for epidemiological parameters Γ, and how it leads naturally to predictions with quantified uncertainty and to anomaly detection schemes. Given probabilistic information on past cases and/or parameters, the probability for new cases is given as

$$P\left[\Delta C(t)\right] = \int d\Delta C(t - \tau) \, d\Gamma \;\; P\left[\Delta C(t) | \Delta C(t - \tau), \Gamma\right] \; P\left[\Delta C(t - \tau), \Gamma\right]. \tag{2}$$

If parameters or previous cases are known exactly, then the distributions under the integral on the right become δ-functions, enforcing specific (observed) values. This identity gives rise to practical schemes for parameter estimation and new case prediction, as we illustrate below.

The probabilistic prescription of new cases (1) implies in turn the distribution of model parameters Γ. The key to parameter estimation, given a probabilistic disease dynamics model, is the well known Bayes' theorem

$$P\left[\Gamma | \Delta C(t) \leftarrow \Delta C(t - \tau)\right] = \frac{P\left[\Delta C(t) \leftarrow \Delta C(t - \tau) | \Gamma\right] \; P\left[\Gamma\right]}{P\left[\Delta C(t) \leftarrow \Delta C(t - \tau)\right]} \tag{3}$$

where the denominator is a normalization factor.

Thus observation of time series for epidemiological quantities ΔC at consecutive times is equivalent to knowledge of the probability distribution for epidemiological parameters. The latter is given, in analogy to (2) as

$$P\left[\Gamma\right] = \int d\Delta C(t-\tau)\, d\Delta C(t)\, P\left[\Gamma | \Delta C(t), \Delta C(t-\tau)\right] P\left[\Delta C(t-\tau), \Delta C(t)\right].$$
(4)

This expression also allows for the inclusion of uncertainties in case counts, which may result e.g. from underreporting or other sampling biases.

3 Probabilistic Disease Models and Real Time Estimation of Epidemiological Parameters

In this section, it is shown how standard "mean-field" compartment models of communicable diseases prescribe $P[\Delta C(t)|\Delta C(t - \tau), \Gamma]$. Alternative models have been proposed to achieve the same goal, including the probabilistic reconstruction of putative chains of transmission [8,22], and stochastic ensemble methods [25]. The approach that we follow here has however the distinct advantage of simplicity. Furthermore, its implementation requires only a very modest computational effort. Other methods present certain advantages, however, such as the estimation of probable chains of transmission [8,21,22].

Mathematical models predicting the time evolution of the average number of infectious cases, deaths, etc. at a given time are among the most useful and most commonly used descriptions of contagion processes. Classical epidemiological models such as SIR (Susceptible-Infectious-Removed) or SEIR (Susceptible-Exposed-Infectious-Removed) are of this form. Each class or compartment counts the mean number of individuals in a specific epidemiological state and may refer additionally to a geographic location, age or risk group.

We have shown elsewhere [23] that in the absence of sources these models imply a relation between new case (or death) numbers at consecutive times, of the form

$$\langle \Delta C(t+\tau) \rangle = b(\Gamma)\Delta C(t),$$
(5)

where the $\langle ... \rangle$ denotes expectation and $b(\Gamma) = \exp\left[\tau\lambda(\Gamma)\right]$. For example in the case of the SIR and SEIR models we have

$$\lambda_{SIR} = \gamma\left(R-1\right); \qquad \lambda_{SEIR} = \frac{\kappa+\gamma}{2}\left[-1+\sqrt{1+\frac{\kappa\gamma}{(\kappa+\delta)^2}(R-1)}\right].$$
(6)

which are the leading positive eigenvalues characterizing the evolution of case numbers. Here κ^{-1} is the latency period, γ^{-1} is the duration of the infectious period, and R is the effective reproduction number, which is defined as the mean number of infected cases caused by an infectious individual. Other models will result in a different form of λ.

These average relations for future new cases can now be used to define a probabilistic model for new cases as

$$\Delta C(t+\tau) \sim P\left[\Delta C(t+\tau) \leftarrow \Delta C(t)|\Gamma\right] = P\left[b(\Gamma)\Delta C(t)\right],$$
(7)

84   L.M.A. Bettencourt et al.

where $\Delta C(t+\tau)$ is taken as a stochastic variable distributed according to the distribution $P[\langle A\rangle]$, with average $\langle A\rangle$. Note that the "mean field" models do not prescribe $\Delta C(t+\tau)$, only its average. Knowledge of higher correlations, such as the variance in case numbers, can help constrain the form of P among a class of discrete (for cases, deaths) probability density distribution functions. In the absence of additional information a Poisson distribution is the highest entropy (i.e. most general) form. Another common choice is a Negative Binomial, which allows for clumping effects, but requires the specification of a quantity related to the variance of $\Delta C(t+\tau)$.

Depending on knowledge of details of the contagion process, correlations among observables and availability of data different model choices in this class may be appropriate, although an SIR or SEIR forms are often adequate.

There are many circumstances when disease cases, or their proxy variables, do not change purely under contagion dynamics. These additional effects may result from multiple introductions, e.g. from an animal reservoir, as well as from unexpected patterns of human behavior. Introductions create cases in addition to those due to contagion, and may start epidemics if initially $R > 1$. To include these effects, we have developed models [23] that account for infections from a reservoir, in addition to the usual transmission from infectious individuals to others. This results in the modification of (5), due to additive terms,

$$\langle\Delta C(t+\tau)\rangle = \Delta B(t+\tau) + b(\Gamma)\left[\Delta C(t) - \Delta B(t) + \tau\gamma R\psi(t,\tau,\Gamma)\right], \quad (8)$$

where $\psi(t,\tau,\Gamma) = \int_t^{t+\tau} dt'\exp\left[-\lambda\times(t'-t)\right]f_c\Delta B(t')$. Here $\Delta B(t)$ are the new cases due to introductions from the reservoir at time t, and f_c is the probability that such cases are transmissible between humans.

This expression shows that inference of disease transmissibility between humans requires statistical knowledge on the number of introductions. This can

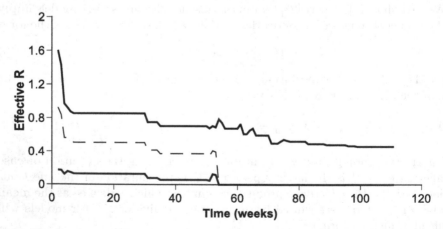

Fig. 2. Predicted effective reproduction number for H5N1 influenza in humans in Vietnam. This is an example of the application of the method described in the text, for details see [23].

usually be inferred from qualitative information and/or can be parameterized to create scenarios, under varying proportions of observable cases being attributable to contagion vs. introductions. We have recently performed such analysis to place bounds on the current transmissibility of H5N1 avian influenza, from data of the outbreaks in Southeast Asia (e.g. see Figure 2)[23].

4 Anomaly Detection

Anomaly detection is not a simple business because often alternative evolution-ary changes may lead to similar outcomes. For example, in the case of communicable diseases these may lead to changes in susceptibility or infectiousness and both "anomalies" often result in comparable outcomes at the population level. Nevertheless, it is desirable to determine when the statistical predictions of epidemiological models fail to describe future data automatically. Inability to predict correctly may signal disease evolution (potentially increasing or decreasing transmissibility or lethality), or changes in the population structure due to population movements and/or unforeseen human behavior.

The probabilistic prediction of new cases (or deaths, etc) that we described above also lends itself to natural and simple schemes for anomaly detection. One approach consists in extracting a prediction for the interval of future cases (or deaths) at a chosen level of confidence (Figure 3). This prescription can be formalized in terms of a two-sided p-value significance test, where $p = \alpha/2$ is the probability that corresponds to a $1 - \alpha$ level of confidence in the predictions of the model. The model is rejected, and an anomaly is flagged, if case numbers are

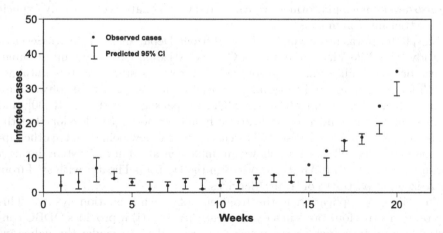

Fig. 3. Schematic illustration of the implementation of anomaly detection. The Bayesian method described in section 3 allows the prediction of the number of cases, with given confidence bounds. Whenever the number of observed cases falls outside that prediction an anomaly is detected and the need for further epidemiological investigation is flagged. Here such events start happening from week 15.

smaller than those corresponding to probability p, or are larger than those corresponding to $1 - p$. The significance α is typically chosen to be $\alpha = 0.05$, which corresponds to (dis)agreement with model predictions at 95% confidence level.

The automatic detection of anomalies should prompt investigations potentially for faults in surveillance, or disease evolution or changes in the population structure. If several measurable quantities are available for prediction and anomaly detection, several statistical tests can be combined into a collective test. The implementation of this aggregated test can take different forms depending on the independency and judged relevance of the individual tests and is the subject of meta-analysis [26]. The changes in disease dynamics (anomalies), may result from co-evolutionary processes, that is, from the interplay between pathogen evolution and changing social landscapes and, consequently, while their detection may be possible, the identification of the underlying mechanisms may not be straightforward without further investigation.

5   Visualization, Interventions and Table Top Exercises

We are currently using the methods described above to create a real-time public health situation awareness and intervention environment. Our objective is to embed real-time statistical inference methods and modeling algorithms in a state-of-the-art visualization and computing environment at the Decision Theater (DT) of Arizona State University. The technology under development will integrate two types of public health applications in a unified user environment: the first is the real-time analysis of geographically tagged public health data; the second is a platform to simulate disease outbreaks for public health planning. As a byproduct, we are also developing a platform for research and visual analysis of spatially-explicit disease transmission models.

The platform was prototyped with the Arizona Department of Health Services to study West Nile Virus in Maricopa County, Arizona, USA. The initial implementation was built using open-source software: a postgres/postgis database [28,27] for the storage and retrieval of spatial data, the statistical software application R [29], algorithms from the DCluster package, written in R [30], and a visualization and rendering application dubbed Minerva [31] developed at the Decision Theater to visualize spatial disease data and now being used in other application areas as well. Figure 4 shows an implementation of the system during a working session to study West Nile Virus in the Decision Theater with staff from the Arizona Department of Health Services.

The Minerva application is the front-end of the information system. This program performs four tasks in a common application: (i) it provides ODBC connectivity to postgres/postgis databases to retrieve data to render through standard SQL, (ii) it renders data in the DT environment using OSG, (iii) it allows users to select layers of data, associated with display properties such as the color and shape, and (iv) it has menus for launching algorithms and choosing parameters, but it does not run them itself. The key to coordinating model runs and data transfers lies with the methods used to access the database. Scripts, written in

Fig. 4. Analysis of West Nile Virus in the Decision Theater at ASU. The photo was taken during a working session at the Decision Theater with the Arizona Department of Health Services, Vector Borne and Zoonotic Disease Department staff. The interactive visualization environment shows spatial analysis of West Nile Virus cases during 2004 and 2005 including demographic and environmental data and spatial clustering algorithms.

R, connect to the database and execute an SQL command that imports a frame of current data into R. Additional R commands transform the data to match the required format for the spatial and statistical analysis. The results are uploaded back into the database into tables which can be queried and rendered by Minerva. Each time a layer is selected for visualization, Minerva starts an R script, waits for the database to indicate it has been updated, and renders the data. Minerva stores binary versions of the rendered layers so they can be turned on and off through the user interface without running the algorithms multiple times during a session.

The fully-developed system will allow the calculation of statistical metrics and mathematical forecasts, coupled with a real-time geographic information system. The power and flexibility of the Decision Theater visualization environment provides the computational and informational infrastructure to deliver a real-time epidemiology decision support system that can be used during public health events, for public health planning, and in developing a better understanding of the spatial spread of multiple diseases. In addition, it allows the visualization of multiple interchangeable data sets simultaneously. Thus, this system can be used for several critical features of a real-time public health information system, such as: 1) data acquisition, storage, and retrieval; 2) exploring the descriptive epidemiology of the disease; 3) determining the major demographic and socio-economic factors that contribute to the spread of the disease, 4) the development of modeling and simulation exercises, and 5) assessing the effects of health interventions in human populations.

The value of using real-time epidemiological decision support environments such as the Decision Theater relies heavily on the access and use by public health officials and on approaches to research and modeling that drive these systems. Current approaches to disease transmission modeling rely too heavily on the analysis of historical data or on the ability to predict future outbreaks, but overlook the needs of public health officials to understand the current unfolding situations in ways that extract meaningful knowledge. The modeling approach in this paper takes a significant step towards the creation of quantitative tools that can deliver real time information to health officials who can best apply the knowledge gained. The visualization of that information in a real-time spatial context is necessary to base real-time decisions that include the current state of disease epidemics as they unfold. Table-top exercises, simulation, and analyses of current infections in the population in cooperation with public health officials also serve to educate health professionals on how to use the information system prior to outbreaks in meaningful ways that will be directly transferable to the monitoring and management of actual health events.

6   Conclusions and Outlook

An integrated quantitative approach to data assimilation, prediction and anomaly detection over real time public health surveillance data streams has been presented. The foundation of the system that we envisage, and its main difference to other current approaches, is the use of epidemiological models as the basis for statistical analyses and temporal prediction. Although these models require more assumptions than purely statistical approaches, they allow for the integration of dynamical variables essential for forecasting and for natural automatic uncertainty quantification and anomaly detection.

The approach developed here will build on the current syndromic surveillance efforts [1,2,3,4], by integrating them with epidemiological predictive modeling, which has a long and successful tradition in support of public health [5,6,7]. While much remains to be done, we hope that these kind of methodologies will enable a shift towards more quantitative active surveillance and primary prevention, resulting in more powerful strategies for monitoring infectious diseases. The integration of these efforts into sustainable solutions that strengthen public health worldwide remains the most important challenge and, simultaneously, the greatest new opportunity to international public health and policy organizations, requiring new levels of transparency, efficiency and cooperation among scientists, governments, the private sector and non-governmental organizations.

Acknowledgments. Portions of this work were done under the auspices of the US Department of Energy under contract DE-AC52-06NA25396. This work was supported in part by grants from the Los Alamos National Laboratory-LDRD (to LMAB), and the National Institutes of Health (RR18754-02, COBRE program from NCRR)(to RMR) as well as by the NSF grant (DMS - 0502349) and NSA grant (H98230-06-1-0097) to CCC.

References

1. Lawson AB, Kleinman K (eds.). Spatial and Syndromic Surveillance for Public Health. Chichester: John Wiley & Sons, 2005.
2. Framework for evaluating public health surveillance systems for early detection of outbreaks; recommendations from the CDC working group. MMWR CDC Surveill. Summ. 2004; 53: 1-16.
3. Proceedings of the 2002 National Syndromic Surveillance Conference. New York, USA, September 23-24, 2002. J. Urban Health. 2003; 80.
4. Various. Abstracts from the 2005 Syndromic Surveillance Conference. Advances in Disease Surveillance. 2006; 1.
5. Anderson RM, May RM. Infectious Diseases of Humans. Oxford: Oxford university Press, 1991.
6. Brauer F, Castillo-Chavez C. Mathematical Models in Population Biology and Epidemiology. New York: Springer-Verlag, 2001.
7. Diekmann O, Heesterbeek JA. Mathematical Epidemiology of Infectious Diseases: model building, analysis and interpretation. Chichester: John Wiley & Sons, 2000.
8. Wallinga J, Teunis P. Different epidemic curves for severe acute respiratory syndrome reveal similar impacts of control measures. Am. J. Epidemiol. 2004; 160: 509-516.
9. Lipsitch M, Cohen T, Cooper B, Robins JM, Ma S, James L et al. Transmission dynamics and control of severe acute respiratory syndrome. Science. 2003; 300: 1966-1970.
10. Riley S, Fraser C, Donnelly CA, et al. Transmission dynamics of the etiological agent of SARS in Hong Kong: impact of public health interventions. Science 2003; 300:1961-1966.
11. Chowell G, Ammon CE, Hengartner NW, Hyman JM. Transmission Dynamics of the Great Influenza Pandemic of 1918 in Geneva, Switzerland: Assessing the Effects of Hypothetical Interventions. J. Theor. Biol. 2006; 241:193-204.
12. Ferguson NM, Donnelly CA, and Anderson RM. Transmission dynamics and epidemiology of dengue: insights from age-stratified sero-prevalence surveys. Phil. Trans. Roy. Soc. Lond. B, 354:757–768, 1999.
13. Koopman JS, Prevots DR, Vaca-Marin MA, Gomez-Dantes H, Zarate-Aquino ML, Longini IM Jr, and Sepulveda-Amor J. Determinants and predictors of dengue infection in Mexico. Am. J. Epidem. 1991; 133:1168–1178.
14. Farrington CP, Whitaker HJ. Estimation of effective reproduction numbers for infectious diseases using serological survey data. Biostatistics 2003; 4:621,632.
15. Hethcote HW. The Mathematics of Infectious Diseases. SIAM Rev. 2000; 42: 599-653.
16. Nishiura H, Schwehm M, Kakehashi M, Eichner M. Transmission potential of primary pneumonic plague: time inhomogeneous evaluation based on historical documents of the transmission network. J. Epidemiol. Community. Health. 2006;60: 640-645.

17. Committee on Modeling Community Containment for Pandemic Influenza. Modeling Community Containment for Pandemic Influenza: A letter report. (National Academies Press, 2006).

18. Castillo-Chavez C, Feng Z, Huang W. On the computation R0 and its role on global stability. In: Castillo-Chavez C, van den Driessche P, Kirschner D, Yakubu A-A (eds.) Mathematical Approaches for Emerging and Reemerging Infectious Diseases: An Introduction. Berlin: Springer-Verlag, 2002: 229-250.

19. Heffernan JM, Smith RJ, Wahl LM. Perspectives on the basic reproductive ratio. Journal of the Royal Society, Interface / the Royal Society. 2005; 2: 281-293.

20. van den Driessche P, Watmough J. Reproduction numbers and sub-threshold endemic equilibria for compartmental models of disease transmission. Math. Biosci. 2002; 180: 29-48.

21. Cauchemez S, Boelle P-Y, Donnelly CA, Ferguson NM, Thomas G, Leung GM et al. Real-time estimates in early detection of SARS. Emerg. Infect. Dis. 2006; 12: 110-113.

22. Cauchemez S, Boelle P-Y, Thomas G, Valleron A-J. Estimating in real time the efficacy of measures to control emerging communicable diseases. Am. J. Epidemiol. 2006; 164: 591-597.

23. Bettencourt LMA, Ribeiro RM. Real time Bayesian estimation of the epidemic potential of emerging infectious diseases. 2007: submitted.

24. Chowell G, Nishiura H, Bettencourt LMA. Comparative estimation of the reproduction number for pandemic influenza from daily case notification data. J. R. Soc. Interface. 2007; 4: 155-166.

25. Bettencourt LMA, Cintron-Arias A, Kaiser DI, Castillo-Chavez C. The power of a good idea: Quantitative modeling of the spread of ideas from epidemiological models Physica A. 2006; 364: 513-536.

26. Petitti DB. Meta Analysis, Decision Analysis and Cost-effectiveness Analysis: Methods for Quantitative Synthesis in Medicine. New York: Oxford University Press, 2000.

27. Website: http://postgis.refractions.net/

28. Website: http://www.postgresql.org/

29. Website: http://www.r-project.org/

30. http://cran.r-project.org/src/contrib/Descriptions/DCluster.html

31. http://sourceforge.net/projects/cadkit

Algorithm Combination for Improved Performance in Biosurveillance Systems

Inbal Yahav and Galit Shmueli

Department of Decision & Information Technologies
and Center for Health Information and Decision Systems
Robert H Smith School of Business
University of Maryland, College Park, MD 20742 U.S.A

The majority of statistical research on detecting disease outbreaks from pre-diagnostic data has focused on tools for modeling background behavior of such data, and for monitoring the data for anomaly detection. Because pre-diagnostic data tends to include explainable patterns such as day-of-week, seasonality, and holiday effects, the monitoring process often calls for a two-step algorithm: first, a preprocessing technique is used for deriving a residual series, and then the residuals are monitored using a classic control chart. Most studies tend to apply a single combination of a pre-processing technique with a particular control chart to a particular type of data. Although the choice of preprocessing technique should be driven by the nature of the non-outbreak data and the choice of the control chart by the nature of the outbreak to be detected, often the nature of both is non-stationary and unclear, and varies considerable across different data series. We therefore take an approach that combines algorithms rather than choosing a single one. In particular, we propose a method for combining multiple preprocessing algorithms and a method for combining multiple control charts, both based on linear-programming. We show preliminary results for combining pre-processing techniques, applied to both simulated and authentic syndromic data.

1 Introduction

Biosurveillance is the practice of monitoring data for the purpose of detecting outbreaks of an epidemic. Traditional biosurveillance has focused on the collection and monitoring of medical and public health data that verify the existence of a disease outbreaks. Examples are laboratory reports and mortality rates. Although such data are the most direct indicators of a disease, they tend to be collected, delivered, and analyzed days, weeks, and even months after the outbreak. By the time this information reaches decision makers it is often too late to treat the infected population or to react in some other way. Modern biosurveillance has therefore adopted the notion of pre-diagnostic ("syndromic") data in order to achieve early detection. Syndromic data include information such as over-the-counter and pharmacy medication sales, calls to nurse hotlines, school absence records, web-searches on medical websites, and chief complaints by individuals who visit hospital emergency rooms. All these do not directly measure

D. Zeng et al. (Eds.): BioSurveillance 2007, LNCS 4506, pp. 91–102, 2007.

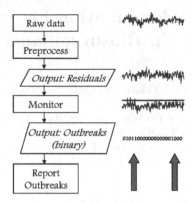

Fig. 1. Analyzing and monitoring data

an infection, but it is assumed that they contain an earlier, though weaker, signature of a disease outbreak. These data are also collected and reported on a much more frequent basis - typically daily. This is one of the many differences in data structure between diagnostic and pre-diagnostic data.

In this work we focus on modern biosurveillance, which is currently implemented in several national-level systems, and that is used to track many series over a large number of geographical areas. We concentrate on temporal monitoring, with a goal of improving the performance of existing monitoring algorithms for automated biosurveillance. In particular, we focus on monitoring univariate pre-diagnostic series. Such series tend to include explainable patterns such as holiday effects, seasonality, and day-of-week effects. For that reason, it is useful to first remove explainable patterns and then monitor the residuals for anomalies that might indicate a disease outbreak [1]. This process is shown in Figure 1.

Most research studies have focused on using a combination of a particular preprocessing procedure with a particular monitoring algorithm. For example, Brillman et. al. (2005) [2] present a combination of square regression with a Shewhart monitor chart. Reis and Mandl (2003) [3] used an autoregressive method integrated with moving average (MA). A literature survey of existing outbreak detection algorithms is outlined by [4,5]. In practice, several of the leading national and regional biosurveillance systems use either no preprocessing or a single simple preprocessing method together with a few different monitoring algorithms (typically Shewhart, MA, CuSum, and EWMA charts). The multiplicity in monitoring algorithms creates multiple testing which results in excessive false alarm rates.

Unlike 'well-behaved' series, where an adequate preprocessing technique can be selected based on the data characteristics, pre-diagnostic data are usually non-stationary, noisy, and extremely heterogeneous across geographies, data sources, and time. In this context it is hard to determine which preprocessing technique is most adequate for each data stream. Furthermore, when the outbreak pattern is known apriori (a signature recognition task), we can design the most efficient monitoring algorithm. However, in the biosurveillance context the nature of a disease outbreak in pre-diagnostic data is typically unknown, and therefore the choice of monitoring algorithm is not straightforward.

In this paper we describe a method for the combination of multiple preprocessing techniques and also of combining a set of monitoring algorithms that outperform any of the single methods in terms of true and false alarms. We take a data-mining approach where the choice of combination is either static or automatically set according to recent performance. To avoid confounding the preprocessing step with the monitoring step, we study each method combination separately: First we combine preprocessing techniques and monitor the combined output with a single monitoring algorithm. Then we use a single preprocessing technique and combine the results from multiple monitoring algorithms.

For combining residuals we use linear combinations of the residuals from multiple preprocessing methods. The weights for the linear combinations are either set statically or chosen adaptively, based on recent performance. The general idea is that some preprocessing function might perform better during some periods (e.g., periods with lower counts), while others might outperform during other types of periods. For combining monitoring algorithms, we combine their binary output (alarm/no-alarm) using experts and majority rules.

The paper is organized as follows: Section 2 introduces the preprocessing methods and monitoring algorithms that we consider. Section 3 describes the proposed algorithms for combining residuals and for combining the output of monitoring algorithms. In Section 4 we describe the simulated and authentic pre-diagnostic datasets and present preliminary results obtained from both . We conclude and discuss future work in Section 5.

2 Preprocessing Techniques and Monitoring Algorithm (Control Charts)

In this section we describe several of the methods used in biosurveillance for removing explainable factors from raw pre-diagnostic data and for monitoring the residuals. These are the building blocks that we later use in the combination methods. These and other preprocessing methods are described in further detail in [1].

2.1 Preprocessing Techniques

We consider one model-driven method and three data-driven techniques:

Linear Regression. Linear and log-linear regression models are a popular method for capturing recurring patterns such as day-of-week, seasonality, and trends [6,2]. The classic assumption is that these patterns do not change over time, and therefore the entire data are used for model estimation.

Differencing. Differencing is the operation of subtracting a previous count from a current one. The order of differencing gives the vicinity between the two counts [7]. We consider 7-day differencing, as suggested by [8].

Moving Average. Moving average (MA) is a technique in which each day estimation is an average of the recent observation. This technique is useful in detecting small process shifts [9].

Holt Winters. Holt-Winters exponential smoothing is a form of smoothing in which a time series is assumed to consist of three components: a level, a trend, and seasonality [10]. [11] show that this is useful for preprocessing pre-diagnostic data.

2.2   Monitoring Algorithms (Control Charts)

The following are the most popular charts in statistical quality control and widely-used in current biosurveillance systems:

Shewhart. The Shewhart chart is the most basic control chart. The daily sample statistic is compared against upper and/or lower control limits (UCL and LCL), and if the limit(s) are exceeded, an alarm is raised. The control limits are typically set as a multiple of standard deviations from the center line [9]. It is most efficient in detecting medium to large spike-type outbreaks.

CuSum. Cumulative-Sum (CuSum) control charts incorporates all the information in a sequence of sample values by plotting the cumulative sums of the deviations of the sample values from the target values. CuSum is known to be efficient in detecting small shifts in the mean of a process [9] and step-function type changes in the mean [12].

EWMA. The Exponentially Weighted Moving Average (EWMA) uses a weighted average of the sample statistics with exponentially decaying weights [13]. It is most efficient in detecting exponential changes in the mean [12].

3   Combination Methods

We consider the problem of *linearly* combining residuals and control chart output for improving the performance of automated biosurveillance systems. In order to better evaluate the separate contribution of the two levels of combination examine preprocessing combinations and monitoring combinations *separately*. When combining residuals from different preprocessing techniques, we use a single control chart. When combining control chart outputs we use *one* preprocessing technique. The two methods are illustrated in Figures 2 and 3.

Our goal is to minimize the rate of false alarms (FA) while minimizing the number of missed alarms (MA). Since this objective is composed of two conflicting sub-goals, we consider the following setup where our goal function is the *cost objective*:

$$\max c = c_{MA}MA + c_{FA}FA$$

where c_{MA} and c_{FA} are nonnegative cost coefficients for MA and FA respectively.

In addition, we distinguish between *static* combination and *adaptive* combination. In the former the optimal linear combination of the methods is determined in advance. In the latter the weights of the function are constantly adapted based on recent performance.

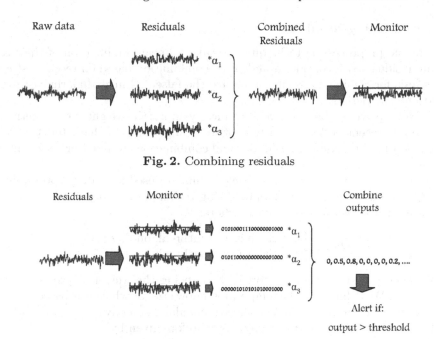

Raw data   Residuals   Combined   Monitor
                                        Residuals

Fig. 2. Combining residuals

Residuals   Monitor   Combine
                                                            outputs

0, 0.5, 0.8, 0, 0, 0, 0, 0.2, ...

Alert if:

output > threshold

Fig. 3. Combining monitor charts

3.1   Combining Control Charts

We assume here that the raw data have undergone a preprocessing step for
removing explainable patterns. Thus, the input into the control charts is a series
of *residuals*. Our goal is to find a *static* linear combination of control charts that
yields the lowest cost value of missed and false alarms. We consider the following
combination rules:

Majority Rule. The algorithm signals an alert at time t only if *more than half*
of the control charts signaled an alarm at time t.

M+n Rule. The algorithm signals an alert at time t only if *all* control charts
in the subset M signaled an alarm, and at least n of the remaining charts
signal an alarm. We use the notation $M+$ if $n = 1$. For example, consider
the set of control charts that were mentioned in Section 2. A *Shewhart+*
algorithm above this set signals an alert only if the Shewhart chart alerts
and at least one other chart (i.e., CUSUM or EWMA) alerts.

The majority rule was first proposed by Condorcet in his 1785 essay. He
considered the case of a group of decision makers reaching a decision on some
issue using the simple majority rule. He made the statement that the group
would be likely to make the correct choice as the size of the group becomes large.
Moreover, the probability of a correct decision increases to 1 as the number of
the *independent* individuals in the group tends to infinity [14].

3.2   Combining Residuals

To combine preprocessing techniques we take a linear combination of their resulting residuals. We consider an *adaptive* combination based on recent history, and search for combinations that minimize the false and missed alarm cost over a certain time period.

In the adaptive combination we dynamically adjust the weight of each method, based on its recent performance, to capture changes in data characteristics. We use a smoothing function of recent optimal combinations to monitor a sequential period.[1]

Formally, we consider k preprocessing techniques used for data preprocessing. Let $R_i[t]$ be the residual from preprocessing technique i on day t. We use O to denote a labeled vector of outbreaks, where:

$$O[t] = \begin{cases} 0 \text{ if there is } \boldsymbol{no} \text{ outbreak on day } t \\ 1 \text{ if there } \boldsymbol{is} \text{ an outbreak on day } t \end{cases}$$

We denote by α a weight vector, with α_i being the weight of preprocessing technique i. We define w as the temporal window size for which α is based on. Finally, T is the alarm threshold, i.e., the system alarms on day t if $\sum_{i=0}^{k} \alpha_i R_i[t] \geq T$. The event of a false alarm on day t is therefore given by

$$FA[j] = 1 \Leftrightarrow \sum_{i=0}^{k} \alpha_i R_i[j] * (1 - O[j]) \geq T \tag{1}$$

and the event of a true alarm on day t is given by

$$MA[j] = 1 \Leftrightarrow \sum_{i=0}^{k} \alpha_i R_i[j] < T * O[j]. \tag{2}$$

Figure 4 illustrates the calculation of the weighted sum of residuals on day t. Our goal is to minimize the cost, given by the sum of FA and MA costs ($c_{FA} \cdot FA + c_{MA} \cdot MA$). With these definitions we can set up the optimization problem as an Integer Programming Problem. Algorithm 1 provides a formalization of this problem.

4   Empirical Study

We describe here some preliminary results that we have obtained by applying the combination methods to simulated and authentic pre-diagnostic data. We start by describing the two datasets and then some initial results.

4.1   Dataset Description

Our data consist of two datasets: simulated data with day-of-week (DOW) effect and authentic emergency department data.

---

[1] Formally, let β be the smoothing coefficient, and α_i be the weight of method i then:
$\forall i, \alpha_i^{new} = \beta \alpha_i^{new} + (1 - \beta) \alpha_i^{old}$.

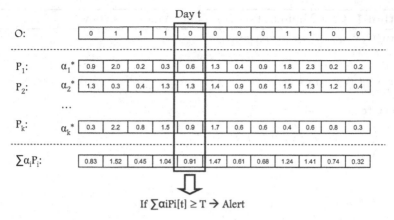

If $\sum \alpha i P i[t] \geq T \rightarrow$ Alert

Fig. 4. Calculating Residuals Weighted Sum for Day j

Simulated Data. The data include 11 generated data series from a Poisson distribution with a DOW effect.[2] Each series include 2000 points (days). The data streams differ in their mean and variance (See Table 1). We used the fifth series to demonstrate the results.

Chief complaints at emergency departments. The authentic dataset includes daily counts of patients arriving at emergency departments in a certain US city, from 2/28/1994 to 12/30/1997 [3]). The counts are grouped into 35 groups by ICD-9 codes. We further grouped the 35 series into 13 categories using the CDC's syndrome groupings [15]. In this work we use the gastrointestinal (GI) category for illustration. Summary statistics for these data are shown in Table 1.

4.2   Outbreak Simulation

To examine the behavior of the combined functions, we simulated two types of outbreaks: a step-increase outbreak with a step size of 0.5σ, and an exponential outbreak of the form $y_t = n \cdot i \cdot e^{\sigma n i}$, where n is the outbreak coefficient and i is the outbreak day. The outbreaks were injected into raw data, i.e. before preprocessing. The outbreaks shapes are illustrated by Figure 5 We varied the outbreak size and length from one experiment to the other. Tables 2 and 3 summarize the characteristics of the simulated step and exponential outbreaks.

---

[2] We thank Sean Murphy from the Johns Hopkins Applied Physics Laboratory for simulating the data.
[3] We thank Howard Burkom of the Johns Hopkins University's Applied Physics Laboratory, for making this aggregated dataset, previously authorized by ESSENCE data providers for public use at the 2005 Syndromic Surveillance Conference Workshop, available to us.

Algorithm 1. Cost Minimization For Time Window [s, s+w]

Input: $\{O, R_1...R_k, T, w, s, c_{FA}, c_{MA}\}$
Output: $\{\alpha\}$

min $c_{FA} * FA + c_{MA} * MA$

Subject to:

$\sum_{i=0}^{k} \alpha_i R_i[t] * (1 - O[t]) - T < FA[t]$
$FA[t] \in 0, 1$
False alarm on day j

$FA = \sum_{j=s}^{s+w} FA[t]$
Total false alarms in window[w, w+s]

$\sum_{i=0}^{k} \alpha_i R_i[t] - T * O[t] < MA[t]$
$MA[t] \in 0, 1$
Missed alarm on day t

$MA = \sum_{j=s}^{s+w} MA[t]$
Total missed alarms in window[w, w+s]

$\sum_{i=1}^{k} \alpha_i = 1$

4.3   Results: Control Chart Combination

We now report preliminary results for combining control charts. Since this work is still in progress, we are yet to have results for combining preprocessing techniques.

We first applied the combination method to the simulated data. We used 7-day differencing to preprocess the data, and then a combination of Shewhart, CuSum, and EWMA charts with different thresholds. We considered five outbreaks of type S1 (Table 2) and five of type E1 (Table 3) *separately*.

The combination rules that we used are: *majority rule*, i.e. at least two monitor charts alert an outbreak, *Shewhart+*, i.e. the method alerts where an alert is generated if the Shewhart chart *and* at least one of the other two charts, *CUSUM+* and *EWMA+*.

Figure 6 shows the ROC curve and the cost value of each control chart separately, and of the combination. For the combined method, we present the top 10 results for different thresholds of the control charts. In both plots we clearly see that a *Shewhart+* method, where the algorithm alerts if the Shewhart chart alerted *and* at least one of CuSum and EWMA charts alerted, gives the best performance (minimum costs: Shewhart+ ≈ 0.49; Shewhart ≈ 0.54; CuSum ≈ 0.57; EWMA ≈ 0.56;).

Table 1. Data Series Characteristics

Simulated Data Series	Mean	Variance
1	0.097	0.16
2	0.53	0.91
3	0.1	1.7
4	3.04	8.67
5	4.96	16.98
6	10.07	55.22
7	20.09	195.05
8	49.87	1084.18
9	99.88	4252
10	500.42	102724.87
11	1000.55	408660.75

ER Data Series	Mean	Variance
Bot_Like	17.05	161.39
Fever	79.18	1747.71
GI	115.75	4350.25
LGI	93.61	2919.04
UGI	22.12	178.95
Hemr_ill	14.79	89.92
Lesion	23.88	149.28
Lymph	6.9	23.89
Neuro	37.38	500.93
Rash	80.96	2182.29
Resp	496.81	77124.22
Shk_Coma	5.45	13.65
UnxplDth	0.58	0.86

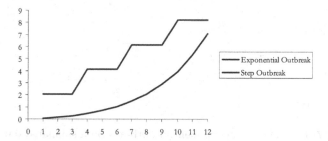

Fig. 5. Outbreaks types

Table 2. Step outbreak types

Name	Length (days)	Step length (days)	Number of Steps
S1	12	3	4
S2	12	5	4

Table 3. Exponential outbreak types

Name	Length (days)	outbreak coefficient (n)
E1	12	0.05
E2	15	0.008

Figure 7 shows the results for the experiments with the exponential outbreak. Here, the Shewhart chart alone yields the best result (true alarm rate ≈ 0.43, cost ≈ 0.72). Yet, for the *same* cost of ≈ 0.39, the best performance is given by the majority rule (≈ 0.73).

Next, we applied the methods to the authentic data. We simulated five outbreaks of type S1 for the first set of experiments, and five of type E2 for the second set.

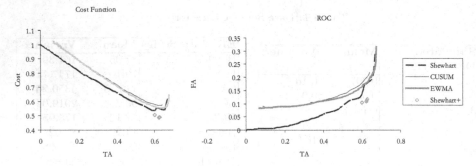

Fig. 6. Determining a step outbreak on a simulated data

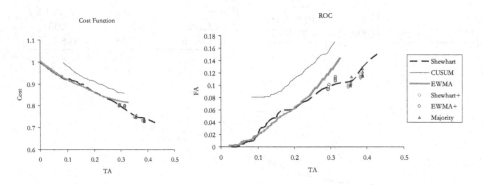

Fig. 7. Determining an exponential outbreak on a simulated data

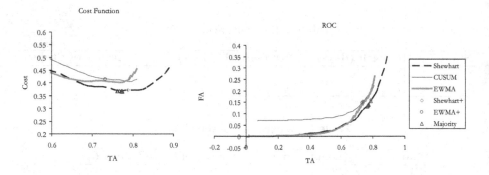

Fig. 8. Determining a step outbreak on the ER data

Figure 8 presents the ROC curve and cost function for detecting the step outbreak. It appears that the majority rule has the best performance (cost of the majority rule ≈ 0.367, whereas Shewhart shows cost of ≈ 0.37). For the exponential outbreak, as presented in Figure 9, we again find that the combination of the three control charts improves the performance by ≈ 0.05 units in favor of Shewhart+.

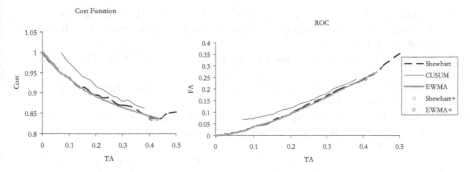

Fig. 9. Determining an exponential outbreak on the ER data

5   Discussion

In this work we proposed a method for improving the performance of univariate monitoring of non-stationary pre-diagnostic data by combining operations at each of the two stage of the outbreak detection task: data preprocessing and residual monitoring. The first consists of combining continuous data (residuals) whereas the second combines binary output (alert/no-alert). By setting an objective cost function that takes into account true and false alarm rates, we are able to formulate this as a linear programming problem and to find the weights that optimize the combination method. Initial empirical experiments confirm the advantage of this portfolio approach. But much more further experimentation is needed.

In the future, in addition to expanding the adaptive combination, we plan to study a machine-learning method that automatically adjusts the combination weights based on current and recent performance, and on the most recent weight vector. The idea is to penalize individual methods whenever the combined method misses a true alarm or detects a false alarm.

Another step is to explore the relationship between linear combinations of control charts and the use of wavelets. It is known that several control charts are related to different resolutions of the Haar wavelet [16]. The extent of this relationship to the combination method is therefore an interesting open question.

References

1. Lotze, T., Murphy, S., Shmueli, G.: Preparing biosurveillance data for classic monitoring. submitted to Advances in Disease Surveillance (2007)
2. Brillman, J.C., Burr, T., Forslund, D., Joyce, E., Picard, R., Umland, E.: Modeling emergency department visit patterns for infectious disease complaints: results and application to disease surveillance. BMC Medical Informatics and Decision Making **5** (2005) 4
3. Reis, B.Y., Mandl, K.D.: Time series modeling for syndromic surveillance. BMC Medical Informatics and Decision Making **3**(2) (2003)
4. Shmueli, G., Fienberg, S.: Current and potential statistical meth- ods for monitoring multiple data streams for bio-surveillance. Statistical Methods in Counter-Terrorism, Eds: A Wilson and D Olwell, Springer (2) (2006)

5. Buckeridgea, D.L., Burkomc, H., Campbelld, M., Hogane, W.R., Mooref, A.W.: Algorithms for rapid outbreak detection: a research synthesis. Journal of Biomedical Informatics **38** (2005) 99–113
6. Rice, J.A.: Mathematical Statistics and Data Analysis. 2 edn. Duxbury Press (1995)
7. Brockwell, P., Davis, R.: Time Series: Theory and Methods. 1991 edn. Springer-Verlag, New York (1991)
8. Muscatello, D.: An adjusted cumulative sum for count data with day-of-week effects: application to influenza-like illness. Presentation at Syndromic Surveillance Conference (2004)
9. Montgomery, D.C.: Introduction to Statistical Quality Control. 3 edn. (1997)
10. Chatfield, C.: The Holt-Winters forecasting procedure. **27**(3) (1978)
11. Burkom, H., Murphy, S., Shmuely, G.: Automated time series forecasting for biosurveillance. Statistics in Medicine (2007)
12. Box, G., Luceno, A.: Statistical Control: By Monitoring and Feedback Adjustment. 1 edn. Wiley-Interscience (1997)
13. NIST/SEMATECH: (e-handbook of statistical methods, http://www.itl.nist.gov/div898/handbook/)
14. Sapir, L.: The optimality of the expert and majority rules under exponentially distributed competence. Theory and Decision **45** (1998) 19–36
15. CDC: Centers for disease control and prevention. (http://www.bt.cdc.gov/surveillance/syndromedef/)
16. Aradhye, H., Bakshi, B., Strauss, R., Davis, J.: Multiscale statistical process control using wavelets - theoretical analysis and properties. AIChE Journal **49** (2003) 939–958

Decoupling Temporal Aberration Detection Algorithms for Enhanced Biosurveillance

Sean Murphy and Howard Burkom

Johns Hopkins University Applied Physics Laboratory, 11100 Johns Hopkins Road,
Laurel MD 20723, USA
{Sean.Murphy, Howard.Burkom}@jhuapl.edu

Abstract. This study decomposes existing temporal aberration detection algorithms into two, sequential stages and investigates the individual impact of each stage on outbreak detection performance. The data forecasting stage (stage 1) generates a prediction of the value of the time series a certain number of time steps in the future based on historical data. The anomaly measure stage (stage 2) compares one or more features of this prediction to the actual time series to compute a measure of the potential anomaly. This decomposition was found not only to yield valuable insight into the effects of the aberration detection algorithms but also to produce novel combinations of data forecasters and anomaly measures with enhanced detection performance.

Keywords: biosurveillance, anomaly detection, time series.

1 Introduction

Multiple steps compose the traditional automated syndromic surveillance system. In the first stage, which we denote as preprocessing or preconditioning, routinely collected sets of medical records such as emergency room visits or over-the-counter medication sales are filtered and aggregated to form daily counts, proportions, weekly aggregates, intervals between cases, and others based on previously defined syndromic classifications. These classifications are generally fixed but may also be created or altered dynamically in response to recent public health information. In the second step, aberration detection, algorithms are used to detect temporal and/or spatial changes in frequencies of the preprocessed data that may be indicative of a disease outbreak. In the third step, the response, syndromic surveillance system users manage alerts by seeking corroboration across time, space and sources of evidence and by initiating the appropriate public health response.

Over the past ten years, various researchers have developed numerous algorithms for the detection of aberrations (step 2) in univariate time series, drawing from such diverse fields as statistical process control, radar signal processing, and finance. The C1 and C2 algorithms of the CDC's Early Aberration Reporting System's (EARS) are based on the Xbar control chart [1], and C3 is related to the CUSUM chart. Reis et al have used a trimmed-mean seasonal model fit with an Auto-Regressive Moving Average coupled to multi-day filters [2], [3]. Brillman et al have used regression based models [4]. Naus and Wallenstein examined the generalized likelihood ratio

D. Zeng et al. (Eds.): BioSurveillance 2007, LNCS 4506, pp. 103–113, 2007.
© Springer-Verlag Berlin Heidelberg 2007

test (GLRT) applied to time series [5]. Burkom et al have examined control charts based on a generalized exponential smoothing implementation of Holt and Winters [6], [7].

As the number of aberration detection algorithms has increased, the situation for both researchers and biosurveillance practitioners has become unclear at best. Despite the fact that some of these algorithms are deployed in operational biosurveillance systems, no general consensus exists as to which is most effective at detecting potential outbreaks.   The aberration detection algorithms continue to increase in complexity. Some algorithms are available for download, while details of others are unobtainable for proprietary reasons. It is resource-intensive for individual researchers to evaluate the performance such algorithms, even when available, on their own data. Exacerbating this situation is the diversity of syndromic time series. Some series have a median count of zero while others have a median in the hundreds. Some exhibit complex cyclic behaviors showing day-of-week (DOW) effects and seasonal fluctuations. For the time series derived from many data sources such as hospital groups or pharmacy chains, the statistical properties of the series change significantly over the course of months and even weeks as the participation by large-scale data providers increases or decreases and as data processing systems and networks improve.   Finally, these algorithms are being applied to improve the detection of disease outbreaks that vary widely in their likely effects on the monitored data. Thus, the unknown shape of the resulting time series signal adds another layer of complexity to the problem. With numerous algorithms, data types and outbreak shapes, two very important questions remain.   Which algorithm works best on a specific type of data and outbreak, and why?

To help answer this question and allow a far more detailed analysis, we separate the aberration detection process into its forecast and anomaly measure stages, denoted below as Stage 1 and Stage 2. While the forecast stage is needed for a measure of data behavior expected in the absence of an outbreak signal, the anomaly stage provides a test statistic based on the difference between observed and expected data. Both stages are not always explicit and may be abbreviated or omitted in some alerting algorithms.

1.1   Stage 1: Data Forecast

The data forecasting stage uses some historic portion of the time series to make n-sample predictions into the future of the expected number of counts. Stage one can be simple or complex. The simplest forecaster would simply use the previous day's value or the value from a week ago without further modification. The EARS family of algorithms—C1, C2 and C3—all use a 7-day moving average to generate the expected values. The only difference among forecasts in these techniques is that C1 uses the last seven days of data while C2 and C3 use data from 3 to 9 days in the past, employing a two-day guardband to help prevent outbreak effects from contaminating the predictions. The g-scan implementation in [5] used a spline fit of historic data for estimation purposes.

Data forecasters can also be adaptive, updating their procedures based on recent data. The Holt-Winters exponential smoother makes predictions using three components: a simple exponential smoother, an adjustment for trends and a cyclic multiplier to handle repetitive patterns. Brillman et al fits an ordinary least-squares, loglinear model to a set of training series values to obtain regression coefficients. These coefficients are used to extrapolate for expected values beyond the training data. Regardless of the complexity of the forecaster, one of the fundamental assumptions in this stage is that the disease outbreaks cannot be predicted *a priori*. I n other words, it is assumed that the predicted series will not contain the outbreak signal and that the appropriate comparison to the actual data will allow timely outbreak identification. For detection of gradual signals, this assumption may necessitate the use of a temporal guardband between the baseline period and the interval to be tested.

1.2 Stage 2: Anomaly Measurement

The anomaly measurement stage compares the observed and forecast values in some fashion to facilitate the decision of whether to investigate a public health event such as an outbreak of infectious disease that may account for the difference between observation and prediction. Typically, the comparison between observed and forecast values is a simple differencing, resulting in a time series of residuals that are then used to form a test statistic as the basis for an alerting decision. Many anomaly measures normalize the residuals either explicitly or implicitly to account for natural variability in the baseline data. The Z-score, related to the Xbar chart of quality control statistics, performs this normalization explicitly by first subtracting a forecast mean from the current observation and then dividing by a standard deviation estimate. This value then serves as the test statistic. The seven-day filtering method employed by Reis does not normalize explicitly but attempts to account for global variability with the determination of an empirical alerting threshold. Often, statistical process control charts such as the cumulative summation (CuSum) or the exponentially weighted moving average (EWMA) chart serve as the anomaly measure stage.

1.3 Motivation

Decoupling the algorithms affords several advantages. Table 1 shows the variety of stage 1 and stage 2 implementations in existing algorithms. Many combinations of these implementations are possible, but few have been explored. The distinction between forecast and anomaly measure becomes blurred once an algorithm becomes associated with a specific organization or publication or acronym, so that alternatives may not be pursued. However, coupling different data forecasters to different anomaly measures can yield novel, potentially enhanced algorithms. Separate consideration of these stages may suggest new approaches or variations to explore and may also help categorize existing techniques and, importantly, understand performance differences among them. Either Stage 1 or stage 2 can be fixed while varying the other stage for testing on a certain type of data and/or outbreak effects.

This study decomposes several popular aberration detection algorithms—an adaptive regression, a Holt-Winters exponential smoother, variations of the EARS family of algorithms, and a temporal scan statistic—into two stages. We then evaluate

Table 1. Decomposition of several existing aberration detection algorithms from the literature into data forecasting and anomaly measure stages

Algorithm	Data Forecasting	Anomaly Measure	Threshold
Fixed Data Threshold	Not applicable	Not applicable	Predetermined number of counts
C1 (EARS)	Moving Average with 7-Day Window	Z-score	3 deviations above the mean (7-day)
C2 (EARS)	Moving average with 7-day window and 2-day guard band	Z-score	3 deviations above the mean (7-day)
C3 (EARS)	Moving average with 7-day window and 2-day guard band	Z-score	3 deviations above the mean (7-day)
Reis [3]	Auto regressive moving average applied to trimmed seasonal model	7-day filter applied to residuals	Empirically derived based on desired sensitivity
GScan [5]	Spline fit to historic data	GLRT applied to fixed summed window	Empirically derived based on simulation studies
Brillman [4]	Loglinear regression model with a fixed baseline	Page's test applied to the residuals in log space	Empirically derived

the effectiveness of all possible aberration detection algorithms assembled from the different combinations of the various stages for detecting two different types of stochastically generated outbreaks inserted into authentic syndromic time series.

2   Methods

The goal of our study was to decompose existing algorithms into two separate stages and evaluate the outbreak detection performance of different combinations of these stages. The outbreak detection capabilities of the various stage-1, stage-2 combinations were evaluated on real syndromic data with stochastically generated outbreak signals inserted in repeated Monte Carlo simulation runs.

2.1   Background Data

The background data for this study were time series of aggregated, de-identified counts of health indicators derived from the BioALIRT program conducted by the U.S. Defense Advanced Research Projects Agency (DARPA) [8]. The appropriate formal agreements to use these data were signed by the authors. Others wishing access may contact the corresponding author for the required procedures. This data set contains three types of daily syndromic counts: military clinic visit diagnoses, filled military prescriptions, and civilian physician office visits. These records, gathered from ten U.S. metropolitan areas, were categorized as Respiratory (RESP),

Gastrointestinal (GI), or Other. While 30 time series were available, 14 series were excluded as they contained artifacts such as temporary dropouts and permanent step increases not representative of routine consumer behavior or disease trends. The remaining data included 10 time series of RESP counts and 6 time series of GI counts, each 700 days in length. All series demonstrated strong DOW effects. The RESP series also demonstrated annual fluctuations, peaking during the winter season, while the GI series did not.

2.2 Stochastic Injects

For the signal to be detected, injected cases attributable to a presumed outbreak were added to the background data. These data epicurves were stochastically drawn from two different lognormal distributions chosen to represent the two extreme types of potential outbreaks. The first, representing a sudden one to two-day jump in cases (Spike), used lognormal parameters $\zeta = 1$ and a $\sigma = 0.1$—see Sartwell et al. for details [9]. The second outbreak type, representing a more gradual rise in the number of cases over time (SlowRise), used $\zeta = 2.4$ and $\sigma = 0.3$. The stochastic epicurves were drawn from the resulting lognormal distribution. To challenge the algorithms, the total number of cases, N, of the outbreak was set to 1, 2, 3, or 4 times the standard deviation of the 4 weeks of detrended background data immediately preceding the start of the injected outbreak. Individual incubation periods were then chosen with a set of N random lognormal draws and rounded to the nearest day. The number of cases to add for each day after onset was then the number of draws that were rounded to that day.

The evaluation process was to add one of the stochastic epidemic curves to the background time series at a randomly chosen start day beyond an 8-week startup period for the alerting method. Each data forecaster was then run on the time series resulting from this linear combination of authentic data and stochastic inject. Residuals were computed from each generated time series of predictions per data forecaster and then run through each anomaly measure, producing a time series of alerting values. This process was repeated for 600 trials for each syndromic time series and outbreak size. As the precise start and stop of each outbreak were known, receiver operator characteristic (ROC) curves were generated for analysis showing the relationship between the probability of detection (y-axis) versus the probability of a false alarm (x-axis) for a range of practical detection threshold values. The same set of epidemic curves was used for testing each data forecaster/anomaly measure pair to minimize the variance of the findings, a technique known as common random numbers.

2.3 Decoupling Aberration Detection Algorithms

Six different data models were used to generate n-day ahead predictions of the time series with and without injects. The first data model used was a generalized exponential smoother based on the Holt Winters method detailed in [10]. The three smoothing coefficients, α, β and γ, were fixed at 0.4, 0 and 0.15 respectively. The second model used was an adaptive regression model using a sliding, 56-day window and covariates for holidays, day of week and linear trend [11]. The third, fourth and fifth models used moving averages with window lengths of 7, 28 and 56 days respectively to make n-day-ahead predictions. The final model used the average of the last 7 weekdays or weekend days to predict n days ahead determined by whether the predicted day was a weekday or weekend [12]. For the detection of spike outbreaks, n was set to two days while, for the slow-rise outbreaks, n was set to 7 days.

We used 4 different anomaly measures to compare the predictions of the data models with the actual time series to compute the likelihood of an anomaly. The first method computed a Z-score based on the prediction residuals (actual count minus predicted count) with the mean of the previous 28 days of residuals subtracted from the current estimate and normalized by the standard deviation of the last 28 residuals. The second method, the weekend/weekday (WEWD) Z-score, used the same method as above with one difference. When computing the output for a weekday, we restricted residuals to the last 28 weekdays. Similarly, when computing the output for a prediction of a weekend day, only weekend residuals were used. Holidays were categorized with weekends. The third method was the fast initial response variation of a cumulative summation with k set to 1 [13]. When the sum exceeded a reset value of 4, the sum was set back to 2 to retain rapid response capability as in [13]. The fourth implemented the g-scan statistic, $G_t(w)$, described by Wallenstein with a 7-day window, w, shown in equation (1) [5].

$$G_t(w) = Y_t(w)\ln[Y_t(w)/E_t(w)] - [Y_t(w) - E_t(w)] \tag{1}$$

$Y_t(w)$ represents the number of counts in the time series within window w and $E_t(w)$ captures the expected number of counts within window w using the predictions provided by the data forecaster.

2.4 Performance Measurement

The six data models coupled to four anomaly detectors yielded 24 aberration detection algorithms. Each of the 24 permutations was tested on 16 different background time series with 2 different inject types (spike and slow-rise) each with 4 different sizes. This resulted in over two thousand ROC curves. To simplify the analysis, the ROC curve information was extracted to form 2 scalar performance metrics: (1) PD84—the probability of detection with a false alarm expected once every 12 weeks and (2) Area—the percentage area under the ROC curve from 1 expected false alarm every 2 weeks to 1 every 16 weeks (see figure 1, panel c).

3 Results

3.1 Sample Results

Figure 1 demonstrates sample output from the two-stage decomposition of aberration detection algorithms along with a ROC curve for one of the RESP time series. Panel (a) shows the output of the data forecasting stage for two different forecasters, a 7-day moving average (MovAvg7, dashed line) and the WEWD 7-day moving average

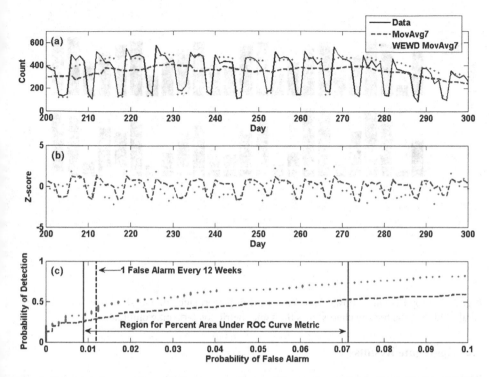

Fig. 1. Panel (a) shows the prediction output of two different data forecasters applied to a RESP time series. Panel (b) demonstrates the output of the Z-score anomaly measure applied to the residuals in (a). Panel (c) illustrates the ROC curve produced by Monte Carlo simulation for the two aberration detections algorithms in (b).

(WEWD MovAvg7, dotted line). The solid black line shows the original time series. Note that the time scale, in days, shows a magnified portion of the 700 day time series. Panel (b) shows the output of the Z-score anomaly measure applied to the moving average 7 residuals (dashed line) and the WEWD 7-day moving average residuals (dotted line) from panel (a).

Panel (c) demonstrates the ROC curves produced with the Monte Carlo stochastic inject simulation using a spike outbreak (1-σ: total injected cases set to 1 standard deviation above the mean) for the two aberration detection algorithms from panel b (using the same line scheme) and graphically illustrates the two performance metrics. PD84 is the y-value at the intersection of the dashed vertical line and the ROC curve and Area is the percentage of area beneath the ROC curve between the two solid vertical lines.

Each of the 24 possible aberration detection algorithms resulting from the different combinations of data forecasters and anomaly measures was applied to the 16 time series as in Figure 1. Figure 2 shows a bar chart that presents the Area results of all tested algorithms for a representative RESP time series. The top chart is for 1-σ slow-rise outbreaks and the bottom chart is for one sigma spike outbreaks.

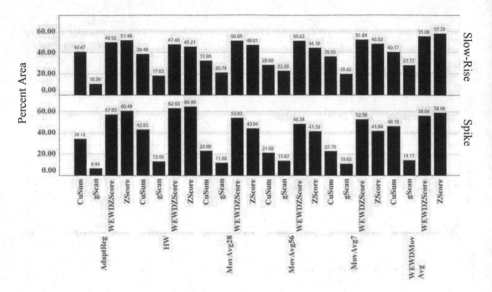

Fig. 2. The top bar chart shows the Area detection performance of 1-σ slow-rise outbreaks for all algorithms. The bottom chart shows the same metric for 1-σ spike outbreaks.

3.2   Aggregate Results

We now turn our attention to aggregate results, averaging the area measures across all 16 data series for the 1-σ outbreak size. Table 2 summarizes these averages for both spike and slow-rise outbreak types. From the table values, it is apparent that a 1-σ outbreak is an extremely challenging test for these aberration detection algorithms. Upon examining these numbers, the large differences in performance between different combinations of data forecasters and anomaly measures become apparent. Looking across rows, we see that the selection of an anomaly measure greatly impacts detection performance. The difference between the highest and lowest performing combinations is as much as 50%. Looking down a column, the performance disparity is smaller, at most 20%.

When examining the relative performance of data forecasters for 1-σ spike outbreaks, one sees that the Holt-Winters method outperforms all other forecasters. The adaptive regression places second when using both Xbar-based anomaly measures, while the weekend/weekday 7-day moving average places second when using the CuSum and g-scan measures. Universally, the data forecasters that adjust for DOW effects outperformed the three simple moving averages. Examining the aggregate performance for spike outbreaks from the perspective of the anomaly measures, the weekend/weekday Z-score consistently outperforms the other three mechanisms. Interestingly, coupling the three data forecasters that adjust for DOW effects with the WEWD Z-score that also adjusts for such variations improved

Table 2. Percent area under the ROC curve averaged across all 16 data series for all possible combinations of data forecasters and anomaly measures for both one sigma spike and 1-σ slow-rise outbreaks

Data Forecasters	Anomaly Measures				
	Z-score	WEWD Z-score	CuSum	G Scan	
Holt-Winters	51.05	51.04	35.61	26.85	Slow-Rise
Adaptive Regression	49.95	50.25	41.04	11.53	
Moving Average 7	52.62	54.25	33.16	22.21	
Moving Average 28	53.27	55.10	33.17	22.44	
Moving Average 56	53.45	53.69	33.35	22.20	
WEWD Moving Average	59.37	58.58	39.08	24.98	
Holt-Winters	59.43	62.03	41.14	15.13	Spike
Adaptive Regression	57.37	57.48	33.74	7.46	
Moving Average 7	39.31	48.18	26.10	10.35	
Moving Average 28	40.59	51.02	26.83	12.97	
Moving Average 56	40.89	49.66	27.40	15.08	
WEWD Moving Average	47.11	51.54	36.65	11.34	

performance over the naïve Z-score. This suggests that the residuals produced by these three data forecasters still contain DOW effects that may bias aberration detection.

For 1-σ slow-rise outbreaks, the situation changes and the weekend/weekday 7-day moving average outperforms all other data forecasters regardless of anomaly detector. From the perspective of the anomaly measures, the Xbar and weekend/weekday Xbar chart perform similarly and outperform the other two measures across data forecasters.

4 Discussion

The aggregate performance results indicate that proper selection of the anomaly measure has a more significant impact on outbreak detection than the data forecasters. The localization of the surge of counts over one or two days from the spike outbreak makes it critically important that either the data forecaster or anomaly measure or both compensate for this effect when detecting such outbreaks in time series with strong of DOW effects. For slow-rise outbreaks that last for longer than a week, DOW compensation does not appear to be as important.

It also appears that some forecasters have a far more detrimental affect on certain control-chart approaches than others. For example, the detection probabilities found with the gscan measure applied to adaptive regression residuals were significantly below those found when it was applied to Holt-Winters residuals. This difference may result from the reduced autocorrelation in the residuals of the latter forecaster, as discussed in [10]. Moreover, while most statistical process control charts typically require independent and identically distributed input data, the detection performance penalty for violation of these conditions needs to be better understood for use with residuals computed for the evolving data streams characteristic of biosurveillance.

A primary systematic feature of the data streams in this study is the day-of-week effect, and algorithms may be classified according to how this effect is modeled. The sliding 7-day average of the widely used C1-C3 ignores this effect, treating all days as equivalent; a count on a Sunday is equivalent to a count on a Monday. For a second management strategy, the CDC "W2" algorithm divides the data into weekday and weekend-plus-holiday counts. The most detailed strategy handles each day uniquely. Both Muscatello's 7-day differencing technique [14] and the regression of Brillman et al with day-of-week indicator covariates fall in this category.

This categorization suggests numerous variations of pre-existing techniques. Treating weekends and weekdays separately appeared to improve the Z-score anomaly measure producing the WEWD Z-Score. It is also possible to modify the other anomaly measures, the CuSum and g-scan, in a similar fashion to handle weekends and weekdays. Furthermore, the 3 base anomaly measures could be stratified by day of week, yielding three additional methods.

5 Conclusion

This study found that the decomposition of aberration detection algorithms into two separate, sequential stages yielded new aberration detection algorithms with improved performance. The analysis of each stage separately offers insight into performance differences among candidate algorithms. This understanding can be used to develop a classification system to select the best aberration detection algorithm to surveil syndromic time series based on data characteristics. For example the current default method used in Electronic Surveillance System for the Early Notification of Community-based Epidemics (ESSENCE) applies a regression model goodness-of-fit measure and automatically selects the algorithm based on the result [15]. When aberration detection algorithms are treated as monolithic techniques, opportunities for substantial improvement are lost.

Furthermore, this decoupling offers numerous research directions. There are many more data forecaster and anomaly measure combinations to test, including simple parameter changes in existing techniques such as the test window length or multiple lengths for the g-scan measure. Also, the anomaly measure variations suggested in the discussion should be evaluated, affording a more precise investigation into the balance between temporal stratification and data adjacency. Forecast techniques from many other fields could also be matched with various control chart ideas. Many novel combinations could be efficiently evaluated with the methodology given here.

References

1. Hutwagner L, Browne T, Seeman GM, Fleischauer AT. Comparing aberration detection methods with simulated data. Emerg Infect Dis [serial on the Internet]. 2005 Feb [date cited]. Available from http://www.cdc.gov/ncidod/EID/vol11no02/04-0587.htm
2. Reis BY, Mandl KD, Time series modeling for syndromic surveillance (2003). BMC Medical Informatics and Decision Making 2003, 3:2.
3. Reis BY, Pagano M, and Mandl KD. Using temporal context to improve biosurveillance. PNAS 100: 1961-1965; published online before print as 10.1073/pnas.0335026100

4. Brillman JC, Burr T, Forslund D, Joyce E, Picard R and Umland E. Modeling emergency department visit patterns for infectious disease complaints: results and application to disease surveillance, BMC Medical Informatics and Decision Making 2005, 5:4, pp 1-14 http://www.biomedcentral.com/content/pdf/1472-6947-5-4.pdf.
5. Wallenstein S, Naus J, Scan Statistics for Temporal Surveillance for Biologic Terrorism. www.cdc.gov/MMWR/preview/mmwrhtml/su5301a17.htm
6. Chatfield C. The Holt-Winters Forecasting Procedure. App Stats 1978; 27: 264-279.
7. Chatfield C and Yar, M. Holt-Winters Forecasting: Some Practical Issues. The Statistician 1988; 37, 129-140.
8. Siegrist D and Pavlin J. BioALIRT biosurveillance detection algorithm evaluation. Syndromic Surveillance: Reports from a National Conference, 2003. MMWR 2004; 53(Suppl): 152-158.
9. Sartwell, PE. The distribution of incubation periods of infectious disease. Am J Hyg 1950; 51:310-318.
10. Burkom HS, Murphy SP, Shmueli G, Automated Time Series Forecasting for Biosurveillance (2007), Statistics in Medicine (accepted for 2007 publication)
11. Burkom, H.S., Development, Adaptation, and Assessment of Alerting Algorithms for Biosurveillance, Johns Hopkins APL Technical Digest 24, 4: 335-342.
12. Tokars J and Bloom S. The Predictive Accuracy of Non-Regression Data Analysis Methods. Poster, 2006 Syndromic Surveillance Conference, Baltimore, MD, October 19-20, 2006.
13. Lucas, J.M. and Crosier R.B. (1984). Fast Initial Response for CuSum Quality-Control Schemes: Give Your CuSum a Head Start. *Technometrics* 24, pp.199-205.
14. Muscatello D, "An adjusted cumulative sum for count data with day-of-week effects: application to influenza-like illness." 2004 Syn Surv Conf, Boston MA, Nov 3-4, 2004.
15. Lombardo JS, Burkom HS, Elbert YA, Magruder SF, Lewis SH et al (2003). "A Systems Overview of the Electronic Surveillance System for the Early Notification of Community-Based Epidemics", Journal of Urban Health, Proceedings of the 2002 National Syndromic Surveillance Conference, Vol. 80, No. 2, Supplement 1, April 2003, New York, NY, p.i32-i42.

Assessing Seasonal Variation in Multisource Surveillance Data: Annual Harmonic Regression

Eric Lofgren, Nina Fefferman, Meena Doshi, and Elena N. Naumova

Department of Public Health and Family Medicine
Tufts University School of Medicine
136 Harrison Ave., 204 Stearns
Eric.Lofgren@tufts.edu

Abstract. A significant proportion of human diseases, spanning the gamut from viral respiratory disease to arthropod-borne macroparasitic infections of the blood, exhibit distinct and stable seasonal patterns of incidence. Traditional statistical methods for the evaluation of seasonal time-series data emphasize the removal of these seasonal variations to be able to examine non-periodic, and therefore unexpected, or 'excess', incidence. Here, the authors present an alternate methodology emphasizing the retention and quantification of exactly these seasonal fluctuations, explicitly examining the changes in severity and timing of the expected seasonal outbreaks over several years. Using a PCR-confirmed Influenza time series as a case study, the authors provide an example of this type of analysis and discuss the potential uses of this method, including the comparison of differing sources of surveillance data. The requirements for statistical and practical validity, and considerations of data collection, reporting and analysis involved in the appropriate applications of the methods proposed are also discussed in detail.

1 Introduction

Seasonal infectious diseases are common and widespread, ranging from viral respiratory viruses such as influenza [1,2] and respiratory syncytial virus [3] to bacterial and vector-borne diseases such as salmonella [4] and west nile virus [5]. They represent significant challenges in both surveillance and control, requiring carefully planned and deliberate epidemiological measurements combined with specialized statistical techniques. We here propose a statistical method for use specifically in the cases of known seasonal periodicity of disease incidence.

Seasonal analysis of time series disease surveillance data has been traditionally confined to the removal of a generic seasonal trend, in order to be able to detect any unusual events: e.g. temporally localized outbreaks or gradual trends over time. While the methods employed for such removal of seasonal patterns have become increasingly sophisticated, they are nevertheless limited by the assumption that "seasonality" exists as a monolithic entity, independent of yearly variation, and capable of being cleanly removed through mathematical means. This may not necessarily be the case. Perhaps more critically, this approach fails to provide new and unique insights into the nature of seasonal diseases that may arise from understanding the observable variation among seasons.

D. Zeng et al. (Eds.): BioSurveillance 2007, LNCS 4506, pp. 114–123, 2007.

Influenza, in particular, has been the subject of a great deal of research incorporating the staggering amount of surveillance data available both in the United States and abroad. The methodology for assessing the "excess" mortality developed by Serfling [6] and colleagues has been used extensively and effectively to analyze large numbers of reported influenza cases. This method has been further refined by Thompson [7,8], Simonsen [9] and others and used for morbidity outcomes. Briefly, "Serfling Regression", as it has come to be called, is a regression model in the form:

$$Y(t) = \beta_0 + \beta_1 t + \beta_2 \cos(2\pi\omega t) + \beta_3 \sin(2\pi\omega t).$$ (1)

Periods where the actual influenza incidence exceeds a threshold 1.64 deviations above this curve are classified as "epidemics". While Serfling Regression has proved invaluable for the study of seasonal influenza, it is incapable of providing any measure of the variability between seasons beyond whether or not the season exceeded this constant epidemic threshold.

Periodic regression models with auto-regressive components as well as Hidden Markov Models provide a better fit compared to a classical Serfling regression and may better follow local seasonal fluctuations but will require a substantial effort to develop interpretable seasonal characteristics.

In this paper, we describe an extension of Poisson harmonic regression model [10], called Annual Harmonic Regression (AHR), which captures seasonal variations between years and allows formal comparison and analysis of factors influencing seasonal characteristics. To illustrate the strengths, limitations, and requirements of the proposed method we apply the technique to PCR-confirmed influenza surveillance data.

2 Annual Harmonic Regression Model

The AHR model fits each season of disease incidence characterized by its own unique curve and is expressed as

$$Y(t)_i = \exp\{\beta_{0,i} + \beta_{1,i} \cos(2\pi\omega t) + \beta_{2,i} \sin(2\pi\omega t) + \varepsilon\}$$ (2)

where $Y(t)_i$ disease incidence at time t within a particular flu season i β_o represents the intercept of the yearly epidemic curve or a baseline level; β_1 and β_2 are the respective coefficients of the harmonic; $\omega = 1/M$, where M is the length of one cycle. Assuming that the errors are uncorrelated, additive, and i.i.d. the standard iterative re-weighted least squares method can be used to estimate the model parameters and their derivatives, and to compute confidence bounds for predicted values. Characteristics of the seasonal curve derived from the regression coefficients: peak timing, intensity, and duration along with their uncertainty measures, are summarized in Table 1 and graphically presented in Figure 1.

Table 1. Annual Harmonic Curve Attributes and Equations

Description	Notation	Expression/Commentary
Variables and outcomes of the regression model		
Time series of disease counts	$Y(t)$	
Time	t	
Length of time series	N	
Length of one cycle	M	$M=52$ for a weekly time series
Time expressed in radians	z	$2\pi\,(t/M)$
Regression parameters intercept, parameter for sin(z), parameter for cos(z).	$\beta_0\ \beta_1\ \beta_2$	
Curve attributes		
Predicted seasonal curve	$S(t)$	
Phase shift - distance of peak from beginning of series expressed in radians	ψ	$-\arctan\{\beta_1/\beta_2\}$
Peak Timing (expressed in week)	P	$M(1-\psi/\pi)\}/2$
Amplitude	γ	$(\beta_1^2+\beta_2^2)^{1/2}$, if $\beta_2>0$; $-(\beta_1^2+\beta_2^2)^{1/2}$, if $\beta_2<0$.
Seasonal peak - maximum value	Smax	$\exp\{\beta_0+\gamma\}$
Seasonal nadir - minimum value	Smin	$\exp\{\beta_0-\gamma\}$
Intensity - the difference between maximum and minimum values	I	$\exp\{\beta_0+\gamma\}-\exp\{\beta_0-\gamma\}$
Relative intensity - the ratio of the maximum and minimum values on the seasonal curve	IR	$\exp\{2\gamma\}$
Seasonal threshold	Sm	$\text{sum}\{S(t)\}/N$, if a seasonal mean is selected for a threshold
Time at which the predicted seasonal curve start to exceed the threshold, week where mean of seasonal curve intersects seasonal curve to the right of peak	T_1	$[\arccos\{(\ln Sm - \beta_0)/\gamma\}-\psi+\pi](M/2\pi)$
Time at which the predicted seasonal curve falls below the threshold, week where mean of seasonal curve intersects seasonal curve to the left of peak	T_2	$52-[\arccos\{(\ln Sm - \beta_0)/\gamma\}+\psi+\pi](52/2\pi)$
Peak duration, number of weeks where seasonal curve is above mean	D	T_1-T_2
Measures of uncertainty		
Standard deviations for the estimates of regression parameters β_1 and β_2, and covariance	$\sigma_{\beta1}$ $\sigma_{\beta2}$ $\sigma_{\beta1\beta2}$	
Variance of Ψ	$\text{Var}(\psi)$	$\{(\beta_1\sigma_{\beta1})^2+(\beta_2\sigma_{\beta2})^2-2\beta_1\beta_2\sigma_{\beta1\beta2}\}/(\beta_1^2+\beta_2^2)^2$
Variance of γ	$\text{Var}(\gamma)$	$\{(\beta_2\sigma_{\beta1})^2+(\beta_1\sigma_{\beta2})^2+2\beta_1\beta_2\sigma_{\beta1\beta2}\}/(\beta_1^2+\beta_2^2)$

Several key attributes of the smoothed curve are graphically displayed below, in Figure 1.

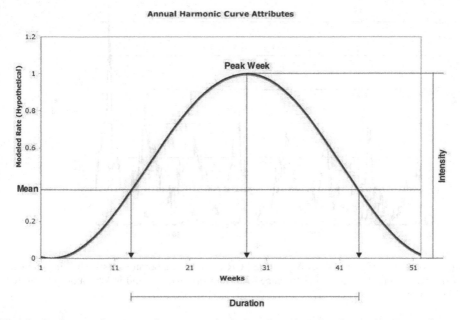

Fig. 1. Attributes of a seasonal curve: peak timing, intensity, duration and a seasonal mean (shifted down with respect to a Poisson process for a theoretical outcome)

3 Case Study: PCR-Confirmed Influenza in Wisconsin, 1996 – 2004

To illustrate the AHR model, 10551 PCR- confirmed tests for influenza virus were examined, of which 910 were positive for influenza virus A, influenza virus B or both. These individual test results were deidentified and abstracted to a weekly time series of total tests and total positive counts over an influenza year, beginning on July 1 and ending on June 30th of the subsequent calendar year (Figure 2). As it represents the bulk of seasonal influenza illness in humans, the analysis was restricted to influenza A, which comprised 636 (70%) of the total positive tests. This time series was chosen for its illustrative purposes, both in the strength and capabilities of the AHR model, and in the requirements for surveillance systems designed to monitor seasonal variation in diseases. This purpose eclipses any particular epidemiological insight into the influenza seasons themselves that may be gleaned from this particular analysis.

The number of positive influenza virus A counts were then fitted using the Poisson AHR model (Figure 3) with the 393 actual weeks of data extended to 419 weeks making up a complete time series over the 8 years. The model describes the influenza A time series with an encouraging degree of accuracy (r^2 =0.622, p<0.001), especially when compared to the traditional Serfling regression (r^2 =0.186, p<0.001).

Total and Influenza Virus A-positive PCR Tests, 1996-2004

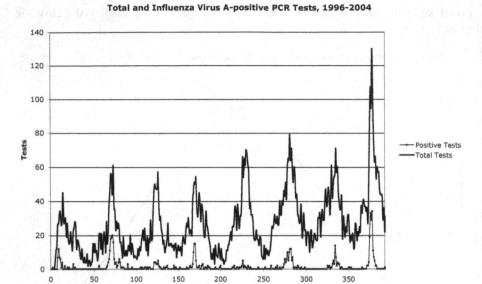

Fig. 2. Total and Influenza Virus A-positive PCR confirmed laboratory tests, showing marked seasonality in both the seasonal occurrence of influenza infection, and in laboratory testing for said infection

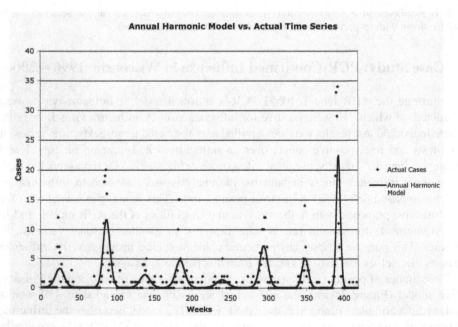

Fig. 3. Annual Harmonic Regression model of influenza virus A-positive PCR-confirmed cases versus the actual time series ($r^2 = 0.622$, p<0.001)

As is apparent in Figure 3, there was considerable annual variation in influenza seasonality, even within the relatively short 8-year time span under consideration. Additionally, within this time frame, the variation between years was seen to be considerably more interesting than was any long-term linear trend in the data (as no such trend was found to be significant). From this model, the key curve attributes were extracted and summarized in Table 2.

Table 2. Annual Harmonic curve attributes for PCR-confirmed influenza A time series

Influenza Year	Peak Week (95% CI)	Intensity (95% CI)	Duration
1996-1997	24.70 (±4.07)	3.222 (±0.767)	18.323
1997-1998	34.23 (±2.85)	11.656 (±0.734)	15.928
1998-1999	33.26 (±4.22)	2.219 (±0.685)	19.756
1999-2000	27.62 (±3.21)	4.661 (±0.574)	18.776
2000-2001	30.36 (±4.71)	1.323 (±0.693)	20.817
2001-2002	32.46 (±3.07)	7.162 (±0.655)	17.32
2002-2003	34.76 (±4.71)	5.059 (±1.336)	15.242
2003-2004	25.38 (±3.65)	18.994 (±1.442)	12.632

The curve parameters obtained for the yearly seasonal increase in influenza are now available for examination and manipulation in a variety of ways that remain inaccessible using only the previously existing time-series methods. These characteristic parameters may be compared internally to examine correlations between measures of timing (Peak Week) and those of severity (Intensity and Duration), as well as over time to reveal patterned shifts in the nature of seasonal increases in incidence over the span of years. In this example, Intensity and Duration were significantly associated ($r=0.886$, $p=0.003$) while Peak Week was not significantly correlated with either Intensity or Durations ($r=0.218$, $p=0.604$ and $r=0.083$, $p=0.844$ respectively). The significant positive correlation is most plausibly due to the use of Intensity in the calculation of Duration.

4 Utility of the AHR Model

Of perhaps greatest use to modern epidemiology and public health is the ability of time series surveillance data, analyzed in this form, to be compared to widely varying data sources. The data from these sources can then be both compared and analyzed in a search for causal linkages. Additional time-series data, such as surveillance data from other sources, mortality records or hospitalization information in the same time scale may be examined to reveal the lag times between an outbreak detected in surveillance systems and the same outbreak detected in hospital records. By using results of AHR model, the time series can be formally compared between differing regions, etiological groups or – with proper preparation – between time-series data presented in different temporal scales.

Beyond the joint-comparison between time-series datasets, surveillance data analyzed using the AHR model may also be analyzed in conjunction with data typically beyond the reach of traditional methods. Annual and non-time-series data, ranging from circulating virus strain, vaccine recommendations, extreme weather events such as El Niño, demographic information, and so on may be used to better understand seasonal influenza. It also allows the examination of disease trends and variation beyond the scope of linear trends captured with traditional techniques that remove seasonality. While the rate of influenza infection in a given period, for example the eight years of the presented study, might not have any pronounced linear trend, there is considerable variation in the timing and severity of individual seasons, which may lead to important insights into the nature of the seasonal periodicity itself that would have gone unnoticed, using previous methods. Importantly, the use of the AHR model allows researchers to examine the effect of a variety of factors using conventional, approachable statistical methods, rather than having to incorporate these variables into a regression model, and allows the analysis of unexpected data that emerges later without forcing a complete re-analysis of the regression model.

There are several benefits to using this technique as well, beyond improving model fit. Through a combination of the curve attributes already described above, and joining the intercepts between adjacent years together to form a quasi-continuous series, the model is able to capture, describe and open for analysis extremely complex, ongoing trends that would be difficult or even impossible to assess using traditional methods without a truly staggering number of linear terms. Finally, the use of year-wise rather than series-wise curves prevents the yearly over- or underestimation of the time series, creating a far less systematic pattern of statistical residuals, leading to the further potential for novel understanding from these patterns.

5 Data Requirements

A consideration must be given to the source of the surveillance data, the intent behind the data collection, and the mechanisms, both practical and theoretical behind the data collection. Seasonality analysis that looks particularly at the variation among seasons, rather than emphasizing the removal of these seasonal trends, must have data that are not seasonally biased by the collection process. In the case study above, there exists an unfortunate and pronounced seasonal cycle not only in influenza incidence, but in the number of tests performed as well. This is understandable, as this particular dataset was not purpose-designed to provide unbiased seasonal surveillance data. Seasonal testing and surveillance bias, while less likely nullify the accuracy of timing estimates, can very much call into question any measure of the severity of a seasonal curve [11]. Alternate sources of time series data, such as mortality records, which posses no such cyclical variation in testing, may provide less biased estimations of severity and timing, although they may still be susceptible to less pronounced bias in the form of physician's perceptions based on the "known" seasonal behavior of some diseases. This reduction in systematic bias comes with a steep price however. Non-specific data sources, such as syndromic surveillance systems or mortality records, do not capture potentially biologically significant information, such as viral subtype in the case of influenza, the presence or absence of particular virulence plasmids in bacterial infections, and lack the specificity of analysis that have made laboratory methods such as PCR and immunoassays the "gold standard" of surveillance data.

It follows that the ideal surveillance system that would allow researchers both to predict and to evaluate seasonal outbreaks, as with traditional Serfling-methodology, as well as attempt to understand the underlying fluctuations in seasonal disease patterns, would combine traits from both laboratory-confirmed and more general surveillance systems. Particularly, every attempt must be made to preserve the accuracy, specificity, and detail of molecular techniques, while designing both protocols and laboratory throughput to prevent cyclical or seasonal bias, regardless of the perceived value of conducting upscaled surveillance for diseases that are "out of season". An example time series of such a hypothetical system, preserving the results used in the real-world example above, is shown in Figure 4. In this scenario, researchers set forth a viral testing schedule and protocol studiously free of cyclical bias, ensuring that any pattern that emerges is due entirely to the seasonality of the disease in question. Beyond merely being unbiased by cyclical testing cycles, the data the emerges from time series studies, especially those using AHR, must use a long-term approach. Many seasons worth of data must be collected for AHR results to approach statistical power, and so it may be preferable in some cases to favor a modest but sustainable study over one that produces spectacular, but short term, results.

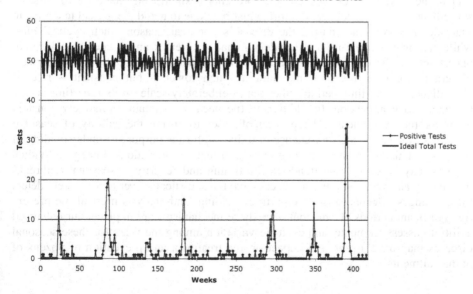

Fig. 4. Hypothetical seasonal surveillance system time series. Time series shows a clear seasonal trend in Influenza incidence, accompanied by a cyclically unbiased collection system, with minor fluctuations in output intended to represent small and inevitable fluctuations in laboratory output.

6 Advancing the Assessment of Disease Seasonality

The role of surveillance in public health and epidemiology has advanced beyond the mere collection and reporting of disease incidence. It has become an essential tool in our ability to detect outbreaks as they emerge, determine their underlying causes, and enact effective interventions. For many diseases, this includes acknowledging and understanding the explicit role of natural, expected seasonal or cyclic fluctuations in incidence. Yet traditional statistical and epidemiological techniques for the study and assessment of seasonality operate from the fundamental assumption that seasonality is constant, and must be removed before truly insightful analysis may occur. For many research questions, this is a perfectly valid, even necessary, approach to questions of seasonality, and existing time-series and seasonal removal techniques unquestionably have a prominent place in the modern epidemiological toolkit. However, furthering our understanding of seasonality also requires examining what variation, if any, exists among seasons, and how these variations themselves might come about.

This type of analysis requires both new statistical techniques and new methods of data collection. Annual Harmonic Regression (AHR) provides the first of these new and necessary methods, giving researchers the means to quantify the characteristics of a specific seasonal outbreak, and to compare seasonal characteristics within a multiyear time series. Additionally, it allows for the cross-comparison of seasonal surveillance data with other data sources, be they other sources of disease incidence information, or non-time series data (e.g. microbiological, genomic, demographic, environmental or otherwise).

This new type of seasonal analysis presents a significant challenge to existing surveillance systems. Deliberate efforts must be made to avoid the logical tendency to scale down surveillance during the disease's non-peak season. Such cyclical bias, while reasonable in terms of laboratory capacity, creates an unacceptable level of uncertainty in the truth of any estimation of seasonal severity. Indeed, it is perhaps preferable to have a system maintain a much lower – but constant – rate of surveillance over a time period rather than deliberately scale up its reporting during the expected peak season. In addition to the need to maintain a relatively constant tenor for the assessment and reporting of cases to ensure the validity of seasonal estimates of severity, the surveillance of a disease for the purpose of understanding of its seasonal nature requires a long-term perspective. Variations among individual seasons may be subtle, and trends in the timing and severity of seasonal outbreaks may take many years before researchers have the statistical power to accurately detect these changes. Mechanisms for reporting, funding, and analysis must all be present for significant periods of time, but once these mechanisms are in place, our enhanced ability to assess, compare, and evaluate variation among the years for these seasonal diseases may have a profound impact on the treatment and prevention of dozens of human ailments.

Acknowledgements

The authors gratefully acknowledge Dr. Kelly Henrickson of the Medical College of Wisconsin for providing the PCR data used as the case study. Additionally, the following funding agencies are thanked for their continual support: The National Institute of Allergy and Infectious Diseases (U19AI062627, HHSN266200500024C) and the National Institute of Environmental Health Sciences (R01ES013171).

References

1. Dowell, S.F., Ho, M.S.: Seasonality of Infectious Diseases and Severe Acute Respiratory Syndrome – What We Don't Know Can Hurt Us. Lancet Infect Dis. **4** (2004) 704-708
2. Lofgren, E., Fefferman, N., Naumov, Y.N., Gorski, J., Naumova, E.N.: Influenza Seasonalty: Underlying Causes and Modeling Theories. J. Virol. (2007) (in press)
3. Graham, N.M.H.. Nelson, K.E., Steinhoff, M.C.: The Epidemiology of Acute Respiratory Infections. In: Nelson, K.E., Williams, C.F.M. (eds.): Infectious Disease Epidemiology: Theory and Practice. Second Edition. Jones and Bartlett, Boston (2007) 699-755
4. Naumova, E.N., Jagai, J., Matyas, B., DeMaria, A., MacNeill, I.B., Griffiths, J.K.: Seasonality in Six Enterically Transmitted Diseases and Ambient Temperatures. Epidemiol Infect. (2007) (in press)
5. Wonham, M.J., de-Camino-Beck, T., Lewis, M.A.: An Epidemiological Model for West Nile Virus: Invasion Analysis and Control Applications. Proc Royal Soc B. **271** (2004) 501-507
6. Serfling, R.E.: Methods for Current Statistical Analysis of Excess Pneumonia-Influenza Deaths. Public Health Rep. **78** (1963) 494-506
7. Thompson, W.W., Comanor, L., Shay, D.K.: Epidemiology of Seasonal Influenza: Use of Surveillance Data and Statistical Models to Estimate the Burden of Disease. J Infec Dis. **194** (2006) 582-591
8. Thompson, W.W., Shay, D.K. Weintraub, E., Brammer, L., Cox, N., Anderson, L.J., Fukuda, K.: Mortality Associated with Influenza and Respiratory Syncytial Virus in the United States. JAMA **289** (2003) 179-186
9. Simonsen, L. Reichert, T.A., Viboud, C., Blackwelder, W.C., Taylor, R.J., Miller, M.A.: Impact of Influenza Vaccination on Seasonal Mortality in the US Elderly Population. Arch Intern Med **165** (2005) 265-272
10. Naumova, E.N., MacNeill, I.B.: Seasonality assessment for biosurveillance systems. In: Advances in Statistical Methods for the Health Sciences: Applications to Cancer and AIDS Studies, Genome Sequence Analysis, and Survival Analysis. Edited by N. Balakrishnan, Jean-Louis Auget, M. Mesbah, Geert Molenberg. 2006. Birkhauser, Boston.(pp. 437-450)
11. Farrington, C.P., Andrews, N.J., Beale, A.D., Catchpole, M.A.; A Statistical Algorithm for the Early Detection of Outbreaks of Infectious Disease. J R Statist Soc A **159** (1996) 547-563

A Study into Detection of Bio-Events in Multiple Streams of Surveillance Data

Josep Roure, Artur Dubrawski, and Jeff Schneider

The Auton Lab, Carnegie Mellon University, Pittsburgh, PA, USA

Abstract. This paper reviews the results of a study into combining evidence from multiple streams of surveillance data in order to improve timeliness and specificity of detection of bio-events. In the experiments we used three streams of real food- and agriculture-safety related data that is being routinely collected at slaughter houses across the nation, and which carry mutually complementary information about potential outbreaks of bio-events. The results indicate that: (1) Non-specific aggregation of p-values produced by event detectors set on individual streams of data can lead to superior detection power over that of the individual detectors, and (2) Design of multi-stream detectors tailored to the particular characteristics of the events of interest can further improve timeliness and specificity of detection. In a practical setup, we recommend combining a set of specific multi-stream detectors focused on individual types of predictable and definable scenarios of interest, with non-specific multi-stream detectors, to account for both anticipated and emerging types of bio-events.

1 Introduction

Maintaining the safety of agriculture and food supply is essential to the well-being of people and the economy. U.S. agriculture encompasses over \$1 trillion in economic activity, including more than \$50 billion in exports [4]. The U.S. food and agriculture systems are naturally vulnerable to disease, pest and contamination. That is due to several factors such as the relative ease of spreading communicable livestock and crop diseases, and simply the traditional methods of breeding and caring for livestock and growing crops.

In the experiments presented in this paper we use real food- and agriculture-safety data made available to us by the United States Department of Agriculture in the framework of an ongoing research project. One of the key objectives of that project is to provide USDA food-, animal- and plant-safety analysts with a surveillance tool capable of effectively monitoring multiple streams of heterogeneous data which is routinely collected by the department. Specific objectives include rapid detection of bio-events, improved situational awareness, and greater ability to anticipate and manage emerging threats to agriculture and food supply.

The particular streams of time series extracted for our experiments correspond to the daily counts of condemned and healthy cattle (labeled as set A in the

D. Zeng et al. (Eds.): BioSurveillance 2007, LNCS 4506, pp. 124–133, 2007.

following parts of the paper), counts of positive and negative microbial tests of meat samples (set B), and counts of passed and failed sanitary inspections of slaughter houses (set C), conducted over a period of about 16 consecutive months in one of the Western U.S. states.

Multi-stream surveillance is attractive because it can lend improvements in sensitivity, specificity and timeliness of detection over more common univariate alternatives. Recently, researchers and practitioners in the field have turned their attention to exploiting the benefits of simultaneous tracking of multiple sources of complementary evidence [6,1,5]. The study presented in this paper focuses on evaluating the utility of approaches to multi-stream analysis in the context of a practical application.

2   Methodology

2.1   Temporal Scan

Each of the data streams we consider consists of two time series of counts. One represents the daily count of detects or positives (such as the animals discarded as being susceptible to a certain illness in the case of stream A), while the other represents the daily count of non-detects or negatives (such as healthy animals approved for slaughter in stream A). We apply statistical analysis to each of these streams in order to measure the extent of a possible departure of the counts observed on a given day from their expected normal levels. We then end up with a set of time series of daily p-values, computed independently for each of the individual streams of source data.

There are may ways of computing such p-values. We apply the method of temporal scan using the popular Chi-Square test and Fisher's exact tests of significance [7]. In temporal scan, a time window of interest, d-days wide, is moved along the time axis and at each discrete step (each day in our case), the numbers of detects and non-detects are aggregated across the d days inside the window and, separately, across all of the remaining days outside of the window. As our data exhibits a very strong day of the week effect, we modify the above procedure such that for the counts outside the window, we only consider the same days of the week as the ones inside the window of interest.

The resulting four counts form a 2-by-2 contingency table for the Chi-Square or Fisher's test procedure. We use Fisher's test if any of the component counts is less than 10 and if the sum of four counts is less than 100, otherwise we apply Chi-Square. The more the observed counts of detects and their proportion to non-detects inside the time window differ from the expectation based on the counts aggregated outside, the lower the p-value obtained for this stream of data on that day.

A simple uni-variate surveillance system would then monitor the individual streams of p-values and trigger an alert if one or more of them went below a pre-set threshold, say $\alpha = 0.05$. It would likely produce many false positives and yet still have only modest power because each of the streams is considered in isolation from the others.

2.2   Multi-stream Non-specific Detector

We next address the question of how, for a given time period, the evidence from each of the individual data streams can be combined into a single detector with more power than any of the individuals. In instances where we have a strong model of the relationships between the streams we might work directly on the raw data [2]. Often no such information is available and that is the case of the data we consider in our empirical results below. Therefore, we adopt a more general approach based on combining the p-values into a single detector.

At first glance, one might be tempted to apply a Bonferroni correction and signal an alarm if the smallest p-value passes the corrected test. This corresponds, however, to an aggregation method based on the *Min* function. The "correction" part of the method actually has no effect on AMOC curves (described in the empirical results section) since the threshold for signaling an alarm is varied to produce the curve.

A more common method that makes better use of all data streams is Fisher's method of combining p-values [3]. He observes that since p-values have a uniform distribution under the null hypothesis, the sum of logs of independent p-values have a χ^2 distribution with $2n$ degrees of freedom: $\sum_{i=0}^{n-1} \ln p_i \sim \chi^2_{2n}$. Conceptually, this approach is easier to understand as computing an aggregate statistic which is the product of the p-values. It also turns out that there is an equivalent closed form solution for the combined p-value,

$$k \sum_{i=0}^{n-1} \frac{(-\ln k)^i}{i!}$$

where $k = \prod_{i=0}^{n-1} p_i$.

In order to illustrate the increased detection power of the combination methods, Figure 1 (right graph) shows the rejection regions for two independent p-values separately ($P1 < 0.05$ and $P2 < 0.05$) as well as for their Fisher's combination ($Fisher's < 0.05$). Note that for the sake of simplicity this example deals only with two data streams. The axes in the graph represent two p-values coming from two independent data streams. If we take into account only one of the p-values, for example $P1$, the null hypothesis rejection region is restricted to the left of the line $P1 = 0.05$. In this case, the null hypothesis is rejected only if $P1 < 0.05$ no matter what is the value of $P2$. Whereas, when we use the *Min* aggregation the rejection region is extended and the null hypothesis is rejected when either $P1$ or $P2$ is below 0.05. Fisher's method also adds to the rejection region cases where both p-values are out of but close to their individual rejection regions. In these cases, Fisher's method is able to combine independent weak signals of departure from the null hypothesis and reject it even when neither $P1$ nor $P2$ individually would lead to reject it.

We refer to Fisher's method as a non-specific detector because it is intended to detect any departure from the null distribution. It is useful even when we have no information about the type of outbreak we want to detect. We note that the generality sacrifices statistical power in the case where we have a model of the outbreak.

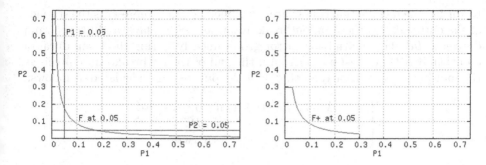

Fig. 1. Left graph: Individuals' and Fisher's aggregated rejection regions. Right graph: Fisher's-based specific rejection region.

2.3 Fisher's-Method-Based Specific Detector

Assume that we have been handed a "scenario" describing a particular kind of outbreak and we would like to improve our detection ability for it. In the empirical tests below, the scenario is one where a ramp up in positive observations occurs simultaneously in three data streams. Note though that the null hypothesis being tested with Fisher's method is that none of the data streams departs from the null distribution.

As a demonstration of the increased power possible from specific detectors, we handcraft an extension to Fisher's method. The modified detector uses the same combined p-value as the original, but then chooses not to signal an alarm if less than two of the uncombined p-values are below a given threshold. Those are cases where the evidence definitely does not match the scenario we are looking for, since only one stream departs from the null distribution.

The graph on the right in Figure 1 shows the rejection region for our example with only two data streams. The additional condition to signal an alarm removes from the Fisher's *non-specific* rejection region areas that do not match our artificial "scenario". The removed areas correspond to those cases where only one of the streams rejects the null hypothesis (low p-value) while the other one does not (large p-value). The effect of the specific detector is to reduce the number of false positives which in turn may allow us to adopt a higher threshold for signaling an alarm and consequently achieve a quicker detection.

3 Experiments and Evaluation

3.1 Artificial Scenario

In order to evaluate the performance of the plain Fisher's method of p-value aggregation and our Fisher-based specific detector we injected artificial outbreaks into the actual data streams. We augmented the actual counts in the streams by using a multiplying factor that linearly ramps up over the period of the simulated outbreak. Such an outbreak is specified with three parameters: (1) the maximum factor by which the actual counts are multiplied, Δ, that represents the strength of the outbreak, (2) the total outbreak duration, Od, in days, and (3) the ramp duration, Rd. The multiplying factor grows linearly during the ramp duration and then it is kept constant at Δ until the end of the outbreak. More precisely, the multiplying factor for the i'th day of the outbreak is calculated as follows:

$$\begin{cases} 1 + (\Delta - 1)/Rd \cdot (i + 1) & \text{if} \quad 0 \le i < Rd \\ \Delta & \text{if} \quad Rd \le i \le Od \\ 1 & \text{otherwise} \end{cases}$$

We injected outbreaks in all of the three data streams simultaneously, beginning at the same day, so that their combination would loosely match the pattern for which our specific detector is designed. The total duration Od was set to 14 days and the ramp duration Rd to 7 days. The Δ parameter was set to 2.0 for stream A, 2.5 for stream B, and 2.3 for stream C. Those parameters were chosen so that the outbreaks would not be immediately detectable in the individual streams during the ramp-up periods.

Figure 2 depicts the time series of p-values for the individual streams of considered data, computed using the 2-day-wide temporal scan window. Note that wider windows render more smoothing and lower the sensitivity of the univariate detectors (for brevity we do not discuss such effects in this paper). The synthetic outbreak can be seen to begin in the second half of October 2005.

It should be noted that it is best to experiment using labeled, known outbreaks identified in the historical data. Unfortunately, at the moment of writing this paper we had not had access to such information, and therefore resorting to realistic but artificial injections was a requirement.

3.2   Min and Fisher's Multi-stream Methods

Figure 3 presents the Activity Monitoring Operating Characteristic (AMOC) curves for the univariate detectors set up for the individual streams A, B and C, and the corresponding curves for the two non-specific multi-stream detectors: *Min* and plain Fisher's (labeled F in the graph). The horizontal axis of the graph corresponds to the number of detects outside of the period of the injected synthetic outbreak, and the vertical axis denotes the time to detection in days from the first day of the outbreak ($i = 0$ in the formula above). The more powerful the detector, the more its characteristic bends towards the lower left corner of the graph. The points and error bars shown are obtained as means and standard errors based on 100 independent injections of simultaneous outbreaks into the individual streams of data, with randomly selected start dates.

It is clear that the non-specific detector implementing Fisher's method of p-value aggregation has a superior detection power over the univariate detectors set on the individual streams of data, as well as over the Bonferroni-correction-motivated *Min* aggregate. This is due to Fisher's method ability to accumulate evidence from independent sources, or data streams, and to include into the rejection region cases where individual streams are close but above the threshold.

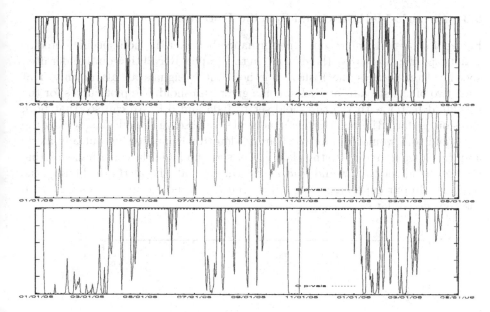

Fig. 2. P-values computed for the individual streams of data using temporal scan with the window width of 2 days. The synthetic outbreak starts on October 15, 2005.

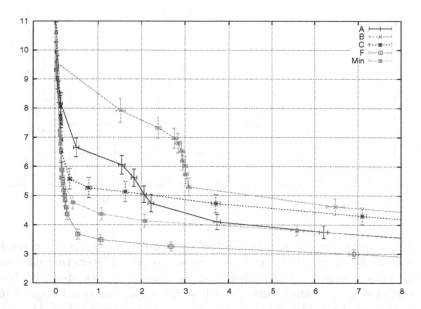

Fig. 3. AMOC curves for univariate detectors (A,B,C) and non-specific multi-stream detectors *Min* and Fisher's

Fisher's method detects the simulated outbreaks faster and at a lower count of potentially false positive detects outside of the scope of the injected outbreaks.

3.3   Fisher's-Method-Based Specific Detector

Figure 4 shows a comparison of performance between the non-specific detector and its specific alternative (labeled F+ in the graph). Recall that our handcrafted specific detector uses the Fisher's method but it signals an alarm only if at least two of the streams are below some given threshold. We plot curves for the specific detector with the threshold set to 0.05 and 0.1. We can see that both cases outperform the plain Fisher's method. We observe though that setting the threshold to 1 would prevent the specific detector from filtering out any alerts and thus it would reduce to the plain Fisher's method. On the other hand, setting the threshold set to 0 would lead to filtering out all the alerts, including true positives. In our experiments we found that a threshold of 0.1 works best with the considered data.

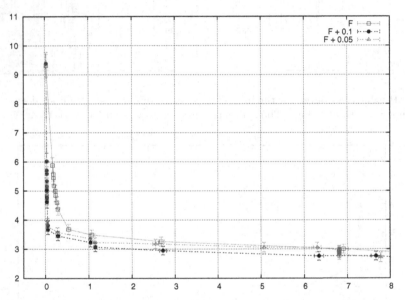

Fig. 4. AMOC curves for multi-stream detectors: the non-specific (Fisher's) and the specific described in Section 2.3 with threshold set to 0.1 and 0.05

We have experimented with other Fisher's-method-based hand-crafted specific multi-stream detectors. There are many potentially useful filters that one could think of. For example, since the considered outbreaks are associated with linear ramp-ups of positive counts during the first few days, one could require that Fisher's-combined p-values for preceding two days are arranged in decreasing order with the current day's p-value, i.e. $p_{day-2} \geq p_{day-1} \geq p_{day0} < \alpha$, in order to signal an alert on the given day. Another possibility could be to rise an alert only if both the current day and the previous day have critical p-values associated with them. That condition would bank on the fact that the outbreaks are known to last several days. Yet another condition could be to require that on

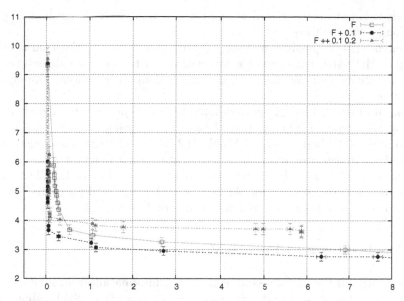

Fig. 5. AMOC curves for multi-stream detectors: the non-specific (F), specific (F+), and specific with a condition on the current and previous day (F++)

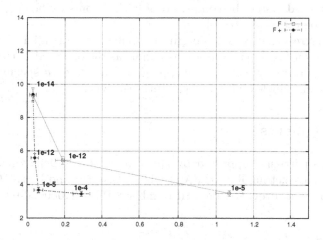

Fig. 6. An improvement of detection time can be achieved using a specific detector in place of its general-purpose counterpart

the current day at least two streams must produce p-values below some threshold and that on the previous day at least one of them was critical. Empirically, none of those ideas were able to outperform the plain Fisher's method. Figure 5 shows the result for the last idea, labeled F++, where the threshold for the current day was set to 0.1 and for the previous day was set to 0.2. We can see that it outperforms Fisher's method until the fourth day of detection latency, but then it also filters out the first days of the outbreak together with false positives and thus it is not able to signal an alert faster than Fisher's method at lower detection latency settings.

Apparently, it is possible to take advantage of knowing the characteristic pattern of the particular type of a bio-event, and to construct detectors with higher specificity and timeliness of detection than their general, non-specific alternative.

Figure 6 illustrates how a specific detector can improve detection time over a non-specific detector. On the plain Fisher curve, using an alert threshold of 1e-12 produces the best result with only 0.2 false positives on average and an average delay in detection of 5.5 days. When the specific detector is added, most of the false positives are eliminated at the alert threshold of 1e-12. Thus for the specific detector, the alert threshold can safely be increased up to 1e-5 or 1e-4 which brings the average time to detect down to 4 days.

4   Conclusions and Future Work

We demonstrated the usefulness of aggregating information from multiple data streams on real food- and agriculture-safety data from the USDA using a general aggregation method. We also showed how a handcrafted aggregation method tuned to specific outbreaks of interest can detect an outbreak faster. We recommend that a fielded system be comprised of both a non-specific detector and as many specific detectors as possible.

We want to stress that manual design of specific detectors might be a difficult task even when outbreak patterns are as simple as the one used in the presented results. In the future work, we will investigate how to learn specific detectors from data streams labeled with various outbreak types.

Acknowledgments

This work was partially supported by the United States Department of Agriculture prime contract number 53-3A94-03-11 (Task 19) and by the Centers for Disease Control and Prevention award number 8-R01-HK000020-02.

References

1. H.S. Burkom, S. Murphy, J. Coberly, and K. Hurt-Mullen. Public health monitoring tools for multiple data streams. *MMWR Morbidity and Mortality Weekly Report*, August 2005.
2. A. Dubrawski, K. Elenberg, A. Moore, and M. Sabhnani. Monitoring food safety by detecting patterns in consumer complaints. In *Proceedings of the National Conference on Artificial Intelligence AAAI/IAAI 2006*, 2006.
3. R. Fisher". *Statistical methods for research workers*. Oliver and Loyd, 1925.

4. Protecting against agroterrorism. GAO-05-214. Technical report, Government Accountability Office, March 2005.
5. D.B. Neill, A.W. Moore, and G.F. Cooper. A multivariate bayesian scan statistic. In *National Syndromic Surveillance Conference 2006*, 2006.
6. M.M. Wagner, A.W. Moore, and R.M. Aryel, editors. *Handbook of Biosurveillance.* Academic Press, 2006.
7. L. Wasserman. *All of Statistics.* Springer, 2004.

A Web-Based System for Infectious Disease Data Integration and Sharing: Evaluating Outcome, Task Performance Efficiency, User Information Satisfaction, and Usability

Paul Jen-Hwa Hu[1], Daniel Zeng[2], Hsinchun Chen[2], Catherine A. Larson[2], and Chunju Tseng[2]

[1] Accounting and Information Systems
David Eccles School of Business, University of Utah
actph@business.utah.edu
[2] Department of Management Information Systems
University of Arizona
{zeng,hchen,cal}@eller.arizona.edu, chunju@email.arizona.edu

Abstract. To better support the surveillance of infectious disease and epidemic outbreaks by public health professionals, we design and implement BioPortal, an advanced Web-based system for cross-jurisdictional information sharing and integration. In this paper, we report two empirical studies that evaluate the outcomes, task performance efficiency, user information satisfaction, and usability associated with BioPortal. Overall, our results suggest that the use of BioPortal can improve users' surveillance performance as measured by analysis accuracy and efficiency (i.e., the amount of time required to complete an analysis task). Our subjects were highly satisfied with the information support of BioPortal and considered it reasonably usable. Our evaluation findings show the effectiveness and value of BioPortal and, at the same time, shed light on several areas where its design can further improve.

Keywords: Infectious disease informatics; public health information systems; cross-jurisdictional information sharing; outbreak detection; system evaluation.

1 Introduction

The surveillance of infectious disease and epidemic outbreaks has become increasingly challenging to public health professionals [1]. Recent epidemic episodes of SARS, Foot-and-Mouth Disease (FMD), West Nile Virus (WNV), and potential outbreaks of avian influenza have attracted extensive media attention, thus creating enormous concerns at both national and international levels [2, 3]. Meanwhile, potential bioterrorism threats are also on the horizon and thereby bring additional complexity to the challenge of surveillance of infectious disease and epidemic outbreaks. As Siegrist [4] noted, bio-chemically competent terrorists can attack people living in a target geographic area by deliberately disseminating infectious diseases using biological agents.

D. Zeng et al. (Eds.): BioSurveillance 2007, LNCS 4506, pp. 134–146, 2007.

Surveillance of an infectious disease or an epidemic outbreak is information intensive and can be greatly supported by effective collection, integration, analysis, and visualization of diverse and voluminous data that are heterogeneous and stored in various sources spanning jurisdictional constituencies horizontally and vertically [5]. A systems-based approach to support important surveillance tasks is appealing. Yasnoff et al. [6] highlight the importance of fruitful collaborations among researchers and practitioners in public health and information systems. We designed and implemented BioPortal [7], a Web-based system that supports convenient access to distributed, cross-jurisdictional health data about several infectious diseases that include WNV, FMD, and Botulism. Specifically, BioPortal supports seamless data integration across different system platforms, contains advanced spatiotemporal data analysis functionalities, and has intuitive, effective visualization capabilities. Preliminary results show encouraging effectiveness of BioPortal, which however needs to be further assessed systematically and methodologically.

In this paper, we report two empirical evaluation studies of BioPortal that focus on outcome, task performance efficiency, user information satisfaction, and system usability. The first study is a controlled experiment that involves 33 graduate students and includes a prevalent spreadsheet-based system for benchmark purposes. The second study is a field evaluation that includes 3 experienced public health professionals affiliated with the State Health Services department in the United States. Overall, our results suggest that the use of BioPortal can improve users' surveillance performance measured by analysis accuracy and efficiency (i.e., the amount of time required to complete an analysis task). Our subjects are highly satisfied with the information support of BioPortal and consider it reasonably usable. Our findings show the effectiveness and value of BioPortal and, at the same time, shed light on several areas where its design can further improve.

2   An Overview of BioPortal

BioPortal is loosely coupled with state public health information systems in California and New York. Each source system transmits WNV and/or botulism related data through secure links to BioPortal using mutually agreed upon protocols. Architecturally, BioPortal is comprised of a Web portal, an internal data store, and a communication backbone. The Web portal provides the necessary user interfaces and allows users to search or query infectious disease–related data sets, visualize these data and analysis results in an intuitive spatiotemporal fashion, perform analysis tasks using built-in analytical models and functions, and identify alerts that signal the development or emergence of "hot spots."

HL7 standards are used as the main storage format to support the necessary data interoperability. All participating agencies transmit data to BioPortal as HL7-compliant XML messages through a secured network connection. HL7 XML-based standards are more advantageous than alternative methods that demand the internal data store to consolidate and maintain the data fields in the respective data sets and sources, hereby offering greater system scalability and extensibility. The communication backbone supports secure data exchanges between BioPortal and each data source. This backbone is built upon widely recognized national standards and provides the necessary modeling and ontological support. Overall, the communication backbone provides data transmission, receiving, and verification functions, together with source-specific data normalization and data security services (using robust encryption technologies). Currently, BioPortal houses several data sets, summarized in Table 1.

Table 1. Infectious disease data sets in BioPortal

Disease	Related data sets
WNV	• Human (NY, CA '03); captive animal (NY '03); • Bird sightings (NY '01-'03, CA '03, USGS '99-'03); • Mosquito pool (NY '03, CA '00); mosquito treatment (CA '04) • Chicken sera (CA '03)
Botulism	• Adult (NY, CA '01-'02); infant botulism (national '04); • Avian botulism (USGS '99-'03)
FMD	• Middle Eastern countries (Iran '87-'03, Iraq '85-'02, Afghanistan '96-'03, Pakistan '85-'03, Turkey '85-'03); South America (Argentina '01)

BioPortal includes spatial temporal visualizer (STV), an advanced visualization module that allows users to explore infectious disease data and examine query results in an intuitive and easily comprehensible manner. STV allows users to load and save spatiotemporal data dynamically for real-time information sharing or further analyses. STV supports several integrated and synchronized views that include periodic, timeline, and GIS. Periodic views enable users to identify prominent periodic temporal patterns. Timeline views provide two-dimensional timelines, together with a hierarchical display of the essential data elements organized in tree structure. GIS views display the reported cases together with their geographic locations on a map.

3   A Controlled Experiment Study

We adopted a randomized, two between-groups design of which the factors are system and general public health knowledge, each defined at two levels (i.e., BioPortal vs. a benchmark system, and low versus high general public health knowledge). Our design supported direct comparisons between BioPortal and the benchmark system and, at the same time, allowed examination of the effect of domain knowledge and its combined impact with the system used. Our subjects were graduate students from the business school or the public health school at a major research university located in the United States. Our subjects participated in the study voluntarily and differed considerably in general public health knowledge; i.e., high for public health students and low for business school students. Each subject was randomly assigned to use BioPortal or the benchmark system, but not both. We were mindful of maintaining a balance in the subject–technology assignment. We administered the experiment to subjects individually or in small groups of two or three.

To assess the outcome associated with the use of a system, we, with the assistance of several experienced public health researchers and professionals, created six analysis scenarios and designed a total of 11 tasks that ranged from simplistic frequency counts to complex trend detection or pattern identification. In Appendix A, we list the analysis scenarios and tasks used in the experiment. We examined task analysis accuracy using a "gold-standard" approach. The experts assisting in the task designs generated a gold-standard analysis result for each task included in the experiment. We measured the accuracy of a subject's analysis of a task on a 10-point scale, with 1 representing "completely incorrect" and 10 denoting "completely correct." We assigned a score of 1 to incomplete tasks. We also assessed task performance efficiency using the amount of time a subject took to complete an analysis task. Our study design administered a 50-minute time constraint, which is considered appropriate according to the results of a pilot study [8]. We explicitly informed each subject of this time constraint before he or she started the experiment tasks.

User satisfaction is fundamental to system evaluation [9]. Specifically, we examined user information satisfaction [10] which emphasizes the user's information requirements. The choice of our focus was made on the basis of the distinct importance of information support to public health professionals. In our study, user information satisfaction refers to the degree to which a user believes a system can satisfactorily meet his or her information needs for an analysis task. We adapted previously validated items to measure user information satisfaction, on the basis of a 7-point Likert scale with 1 being "strongly disagree" and 7 being "strongly agree." We assessed system usability using the QUIS [11], a common instrument that has been widely used in various information systems. The usability of each investigated system is evaluated in terms of a user's overall reaction to the system, his or her assessment of the screen layout and sequence, terminology and system information, system learnability, and system capabilities. Each usability dimension was measured by multiple items, on the basis of a nine-point Likert scale.[1] We used a scripted document to inform all subjects explicitly of the study's purpose, experimental procedure, and our analysis and management of the data to be collected in the experiment. We specifically addressed concerns about information privacy and ensure that we would perform data analyses at an aggregate level, not in any personally identifiable manner.

The hypotheses tested in the experiment are as follow:

H1: The outcome accuracy resulting from the use of BioPortal is significantly greater than that associated with the benchmark system.

H2: The amount of time a subject needs to complete an analysis task is significantly less when supported by BioPortal than by the benchmark system.

H3: The user information satisfaction associated with the use of BioPortal is significantly higher than that observed with the benchmark system.

---

[1] Details of the scale used in QUIS are available in [11]. In general, lower scores represent more favorable usability assessments (e.g., easy, wonderful, clear) than higher scores (e.g., difficult, terrible, confusing), with 1 being the most favorable and 9 being the most unfavorable.

H4: Users are likely to consider BioPortal more usable than the benchmark system and assign higher usability scores for their overall reactions to the system, screen layout and sequence, terminology and system information, system learnability, and system capabilities.

4 A Field Evaluation Study

We conducted a field evaluation that involved experienced public health professionals. The objective of this study was to evaluate BioPortal by involving public health professionals in their work context. With the assistance of several domain experts, we designed analysis scenarios and tasks to mimic the surveillance tasks common to public health professionals. Our field study also focused on analysis accuracy, task performance efficiency, user information satisfaction, and system usability, using the same measurements from the controlled experiment described previously.

We also collected subjects' qualitative assessments of BioPortal, using the following semi-structural questions: "Does BioPortal provide sufficient query criteria or support (e.g., different ways to query)? If not, what additional query criteria or support should be included?"; "How useful are aggregated views? In what particular ways do such views help your performing an analysis or problem solving task?"; "How useful is the GIS tool? How can it be improved to better support your performing analysis or problem-solving tasks?"; and "How useful is the Timeslider? Please list 2 or 3 important ways in which this tool is helpful to your analysis or problem solving." We also used questions adapted from [12] to assess subjects' intentions for using BioPortal in their work contexts. The specific analysis scenarios and tasks used in the field study are listed in Appendix B.

5 Results and Discussion

5.1 Controlled Experiment Results

Our controlled experiment had 33 subjects, 17 using BioPortal and 16 using the spreadsheet system. In the BioPortal group, 9 subjects were public health students and the remaining 8 subjects were from the business school. In the spreadsheet system group, 7 subjects were public health students and the remaining 9 subjects were from the business school. We had 20 male subjects and 13 female subjects; our subjects have comparable demographic characteristics (including age and education) and are similar in general computer efficacy and Internet usage.

Analysis Accuracy: We used the corresponding gold-standard result to evaluate the accuracy of each analysis task performed by each subject. We aggregated the analysis accuracy for a subject across all the experiment tasks he or she performed and used the resulting accuracy to test the hypothesized effect of technology.[2] As we show in Table 2, our analysis demonstrated that technology has a significant effect on analysis accuracy. Specifically, the accuracy associated with BioPortal (mean = 81.94, SD = 21.23) was greater than that of the spreadsheet program (mean = 61.19, SD = 17.92), and the difference was significant at the 0.01 level. Thus, our data supported H1; i.e., the outcome accuracy resulting from the use of BioPortal would be significantly greater than that associated with the benchmark system.

Table 2. Analysis of effects on analysis accuracy

Source	DF	Type III SS	Mean Square	F-Value	P-Value
Domain knowledge	1	1,165.43	1,165.43	3.12	0.08
Technology	1	3,173.12	3,173.12	8.51	0.00
Domain knowledge x Technology	1	28.54	28.54	0.08	0.78

Task Performance Efficiency: We examined task performance efficiency using the amount of time a subject needed to complete an analysis task. As shown in Table 3, technology has a significant effect on the amount of time a subject needed to complete a task (p-value < 0.01). On average, subjects who used BioPortal could complete a task considerably faster (mean = 36.28 minutes, SD = 11.33 minutes) than their counterparts supported by the spreadsheet program (mean = 48.23 minutes, SD = 5.07 minutes); the difference was significant at the 0.01 level.[3] Thus, our data supported H2; i.e., the amount of time a subject needs to complete an analysis task would be significantly less when using BioPortal than using the benchmark system.

Table 3. Analysis of effects on task completion efficiency

Source	DF	Type III SS	Mean Square	F-Value	P-Value
Domain knowledge	1	673,239.59	673,239.59	2.43	0.13
Technology	1	4,395,727.04	4,395,727.04	15.84	0.00
Domain knowledge x Technology	1	344,002.34	344,002.34	1.24	0.27

User Information Satisfaction: According to our analysis, technology has a significant effect on user information satisfaction (p-value < 0.01). As shown in Table 4, subjects using BioPortal exhibited higher satisfaction with the information support

---

[2] When aggregating the analysis accuracy of a subject across experimental tasks, we assigned an accuracy score of 1 to each incomplete task. Our rationale is that an incomplete analysis, from a grading perspective, is not any better than one that is completely incorrect, particularly with regard to the corresponding gold-standard solutions in the assessment.

[3] For a task not completed within the specified time limit, we used 50 minutes as the time requirement. In light of the much lower task completion rate associated with the spreadsheet program, the actual time requirement difference between the investigated systems may be greater than that reported herein (which is already prominent and significant statistically).

(mean = 2.34, SD = 1.02) than their counterparts supported by the spreadsheet program (mean = 3.68, SD = 1.23); the difference was significant at the 0.01 level. Thus, our data supported H3; i.e., the user information satisfaction associated with the use of BioPortal would be significantly higher than that observed with the benchmark system.

Table 4. Analysis of effects on user information satisfaction

Source	DF	Type III SS	Mean Square	F-Value	P-Value
Domain knowledge	1	0.80	0.80	0.68	0.42
Technology	1	13.38	13.38	11.48	0.00
Domain knowledge x Technology	1	5.08	5.08	4.36	0.05

System Usability: According to our analysis, technology has a significant main effect on both overall reactions to the system (*p*-value < 0.01) and system capabilities (*p*-value < 0.05) but not on screen layout and sequence or terminology and system information. The effect on system learnability is somewhat significant statistically, as suggested by a *p*-value between 0.05 and 0.10. Overall, our subjects considered BioPortal generally usable and recognized its utilities for supporting their analysis tasks. However, the evaluation results indicated that the design of BioPortal may need improvement in its screen layout and sequence, as well as its language in terms of clarity and user friendliness. Our subjects considered learning to use BioPortal not particularly difficult, but its learnability could be enhanced further. Overall, our evaluation results suggested that BioPortal arguably is more usable than the spreadsheet program in most, but not all, the fundamental usability dimensions. Thus, our data partially supported H4; i.e., user would be likely to consider BioPortal more usable than the benchmark system and assign higher usability scores for their overall reactions to the system, screen layout and sequence, terminology and system information, system learnability, and system capabilities.

5.2 Field Evaluation Results

A total of three public health professionals took part in our study voluntarily (one female and two males). Our subjects were between 31 and 36 years old and had doctoral degrees in public health (or related disciplines). Our subjects self-reported reasonable general computer efficacy and used the Internet on a frequent and routine basis. Each subject was knowledgeable about epidemiological practice and showed great confidence in analyzing and interpreting data about different infectious diseases or epidemic outbreaks.

As a group, our subjects showed satisfactory analysis accuracy, scoring an average of 1.91 on a 2-point scale (2 = completely correct; 1 = partially correct; 0 = incorrect). On average, the subjects were able to complete all analysis tasks in one hour and twelve minutes (SD = 11.67), which, according to our subjects, was noticeably shorter than that commonly needed to complete their analysis tasks using existing systems. Analysis of our subjects' evaluative responses suggested high user information satisfaction; i.e., mean = 5.78 and SD = 1.12 on a seven-point Likert scale, with 7 being "strong agree." Our subjects exhibited high intentions for using BioPortal in their work context; i.e., mean = 6.0 and SD = 1.24 on a seven-point Likert scale, with 7 being "strong agree." Table 5 summarizes the results of subjects' user information satisfaction and their intentions for using BioPortal in their work contexts.

Table 5. Summary of user information satisfaction and intention for using BioPortal

Measurement item	Mean	S.D.
User Information Satisfaction (UIS)	*5.78*	*1.12*
UIS-1: BioPortal offers valuable utility in my analysis of public health problems or trends.	6.25	0.96
UIS-2: I can understand the functions of BioPortal.	6.25	0.96
UIS-3: Using BioPortal can quickly generate the analysis results that I need.	6.00	0.82
UIS-4: The analysis results by BioPortal are reliable.	4.25	1.53
UIS-5: The visualization designs of BioPortal are good.	6.50	1.00
UIS-6: In general, I am satisfied with the response time of BioPortal.	6.50	0.58
UIS-7: Overall, I find the results generated by BioPortal to be relevant to my analysis of public health problems or trends.	6.25	0.96
UIS-8: The analysis results by BioPortal are accurate.	4.25	1.53
UIS-9: Overall, I have good control over using BioPortal to complete an analysis task.	5.75	0.96
UIS-10: BioPortal is flexible in supporting different analysis tasks in public health.	5.75	1.90
Intention to Use BioPortal	*6.00*	*1.24*
BI-1: When I have access to BioPortal, I would use it as often as needed.	6.00	1.41
BI-2: To the extent possible, I intend to use BioPortal in my job.	6.00	1.15
BI-3: Whenever possible, I would use BioPortal for my tasks.	6.00	1.15

The qualitative assessments of BioPortal were mostly positive. One subject commented, "Capability of BioPortal is huge–could link to state data and would have a great foundation. Serotypes also useful for linking cases epidemiologically." According to our subjects, BioPortal has adequate design and is easy to use. One subject pointed out that "design is one of its strengths – very intuitive and user friendly." Another subject noted that "BioPortal is a little easier to use than existing syndromic surveillance systems." The subjects were particularly fond of the spatial temporal visualizer in BioPortal. As one subject commented, "The GIS thing is big! Especially West Nile virus is big in [our] County – [we] would like to be able to look at the geographic spread. This could influence mosquito intervention, and see movement over time, when cases stop and/or pop up somewhere else. Would also be good for rabies." Similarly, the other two subjects also had positive feedback, commenting that "Just being able to pick a time period (for example, one week) and see how it unfolds. Also seeing the faded out cases very helpful," and that "Aggregated views are good for overall picture of data and for answering specific questions by choosing 2x2 table variables," respectively. Our subjects particularly liked the built-in hot spot analysis in BioPortal, commenting that "Hotspot analysis is instrumental to what this user does everyday – the job function is to detect any health event in the community before diagnoses;" and that "The hotspot analysis tool embedded in STV is very useful. When user clicked a mock data set, it went straight to STV, then used the tool to pick SatScan and parameters. Would be easier to use that way to change baseline."

Our subjects in general considered BioPortal reasonably usable, as manifested by a mean of 2.42 in overall reactions towards the system (SD = 1.40) (Table 6).

Table 6. Summary of BioPortal usability evaluation results

Measurement Item	Mean	S.D.
A. *Overall Reactions towards the System*	*2.42*	*1.40*
(wonderful/terrible)	2.50	1.29
(satisfying/frustrating)	3.25	1.89
(stimulating/dull)	1.75	0.96
(easy/difficult)	2.50	1.73
(adequate utility/inadequate utility)	2.25	1.26
(flexible/rigid)	2.25	1.26
B. *Screen Layout and Sequence of System*	*1.94*	*0.75*
Characters on the computer screen	2.50	0.58
Visual design of the screen simplifies task	1.75	0.50
Organization of information on screen	1.75	0.96
Sequence of screens	1.75	0.96
C. *Terminology and System Information*	*2.00*	*1.16*
Use of terms throughout system	2.00	1.15
Terminology presented on the interface is related to the "analysis task"	2.25	1.26
Position of messages on screen	1.50	0.58
Messages on screen which prompt user for input	1.50	0.58
Computer keeps you informed about what it is doing	2.75	2.22
D. *Learning to use the System*	*3.08*	*2.42*
Learning to operate the system	2.50	2.38
Exploring new features by trial and error	2.75	1.50
Remembering commands or making menu choices for performing searches and new analyses	2.50	2.38
Tasks can be performed in a straight-forward manner	3.25	2.63
System-provided help messages or instructions	4.25	3.40
Readability of system-provided instructions or online help	3.25	2.22
E. *Capabilities of the System*	*1.95*	*1.24*
System speed	1.50	0.58
System reliability	1.25	0.50
System tends to be	2.00	1.41
Correcting mistakes is	1.50	0.58
Experienced and inexperienced users' needs are taken into consideration	3.50	3.11

According to our analysis, BioPortal seems to be more usable in terms of screen layout and sequencing, system capabilities, and terminology and system information than in learning to use the system which represents one where the design of BioPortal needs to improve. In particular, built-in instructions need further improvement, as suggested by the subjects' lukewarm responses to "system-provided help messages or instructions" and "readability of system-provided instructions or online help."

6 Summary

We conducted two empirical studies to evaluate the outcome, task performance efficiency, user information satisfaction, and usability associated with BioPortal. Overall, our results were encouraging and suggested BioPortal can enhance public health professionals' surveillance of infectious disease or epidemic outbreak in terms of analysis accuracy and time requirements. Our findings show subjects exhibiting high user information satisfaction when supported by BioPortal. In addition, BioPortal appeared reasonably usable but its built-in instructions need to improve in order to better guide public health professionals to use the system to complete their analysis tasks. Our future research plans include performing field studies to evaluate the outcome and user impacts of BioPortal in real-world public health settings, and examine its acceptance by public health professionals and researchers using a large-scale survey study.

References

1. C.D. Ericsson and R. Steffen, "Population mobility and infectious disease: The diminishing impact of classical infectious diseases and new approaches for the 21st century," *Clinical Infectious Diseases*, vol. 31, pp. 776-780, 2000.
2. Y. Li, L.T. Yu, P. Xu, J.H. Lee, T.W. Wong, P.L. Ooi, and A.C. Sleigh, "Predicting super spreading events during the 2003 Severe Acute Respiratory Syndrome epidemics in Hong Kong and Singapore," *American Journal of Epidemiology*, vol. 160, pp. 719-728, 2004.
3. S.B. Thacker, A.L. Dannenberg, and D.H. Hamilton, "Epidemic intelligence service of the Centers for Disease Control and Prevention: 50 years of training and service in applied epidemiology," *American Journal of Epidemiology*, vol. 154, pp. 985-992, 2001.
4. D. Siegrist, "The threat of biological attack: Why concern now?" *Emerging Infectious Diseases*, vol. 5, pp. 505-508, 1999.
5. R. Pinner, C. Rebmann, A. Schuchat, and J. Hughes, "Disease surveillance and the academic, clinical, and public health communities," *Emerging Infectious Diseases*, vol. 9, pp. 781-787, 2003.
6. W.A. Yasnoff, J.M. Overhage, B.L. Humphreys, and M.L. LaVenture, "A national agenda for public health informatics," *Journal of American Medical Informatics Association*, vol. 8, pp. 535-545, 2001.
7. D. Zeng, H. Chen, L. Tseng, C. Larson, M. Eidson, I., Gotham, C. Lynch, and M. Ascher, "West Nile virus and botulism portal: A case study in infectious disease informatics," *Lecture Notes in Computer Science*, Vol. 3073, H. Chen, R. Moore, D. Zeng, and J. Leavitt, Eds. Springer, 2004, pp. 28-41.

8. P.J. Hu, D. Zeng, H. Chen, C. Larson, W. Chang, and C. Tseng, "Evaluating an infectious disease information sharing and analysis system," *IEEE International Conference on Intelligence and Security Informatics* (IEEE ISI 2005), Atlanta, GA May 19-20, 2005; *Lecture Notes in Computer Science*, vol. 3495, April 2005.
9. W.H. DeLone and E.R. McLean, "Information systems success: The quest for the dependent variable," *Information Systems Research*, vol. 3 (1), pp.60-95, 1992.
10. B. Ives., M. Olson, and J.J. Baroudi, "The measurement of user information satisfaction," *Communications of the ACM*, vol. 26 (10), pp.785-793, 1983.
11. J.P. Chin, V.A. Diehl, and K.L Norman, "Development of an instrument measuring user satisfaction of the human-computer interface," *Proceedings of the ACM CHI '88*, Washington, DC, 1988, pp.213-218.
12. F.D. Davis, "Perceived usefulness, perceived ease of use, and user acceptance of information technology," *MIS Quarterly*, vol. 13, pp. 319-339, 1989.

Appendix A: Listing of Analysis Scenarios and Tasks Used in the Experiment

Scenario 1: Examine data related to WNV.
- Task 1: In 2002, which county in New York had the highest dead bird count?
- Task 2: Of the three listed bird species, Bluejay, Crow and House Sparrow, which had the highest number of positive cases of West Nile Virus?

Scenario 2: Examine a correlation between Botulism and gender.
- Task 3: In California, for year 2001, did more men or more women suffer from Botulism?
- Task 4: In California, for year 2002, did more men or more women suffer from Botulism?

Scenario 3: Determine the occurrence of Foot-and-Mouth disease in 2001 for three countries.
- Task 5: In 2001, in which week(s) do the highest number of Foot and Mouth Disease cases occur in Iran?
- Task 6: In 2001, in which week(s) do the highest number of Foot and Mouth Disease cases occur in Turkey?
- Task 7: In 2001, in which week(s) do the highest number of Foot and Mouth Disease cases occur in Argentina?

Scenario 4: Determine the location of the most intensive outbreak of WNV during 1999 in New York.
- Task 8: During 1999, where (in which county) and when did the most intensive occurrence (i.e., highest number of cases) of West Nile Virus happen in New York State?

Scenario 5: Describe the spread (geographically and over time) of dead crow sightings for New York in 2002.
- Task 9: Please describe the spread (geographically and over time) of dead crow sightings in New York in 2002.

Scenario 6: Determine correlations between the incidence of WNV and dead bird occurrences and mosquito pool counts.
- Task 10: Using the BioPortal system or the spreadsheets, as assigned, to investigate West Nile Virus disease can you determine if, during 2002, there is a correlation between the dead bird occurrences and mosquito pool counts?
- Task 11: (Continued with Task 10) If so, what correlation do you observe?

Appendix B: Listing of Analysis Scenarios and Tasks Used in the Field Evaluation Study

Scenario 1: BioPortal Website Functionalities
This scenario will focus on the use of the BioPortal website. In this scenario, the user is asked to provide characteristics of the target dataset. These characteristics include:

- the number of cases with certain syndromes within a time period
- the date of the first case of a certain syndrome
- detailed case information

In this scenario, the user will make use of the following BioPortal functionalities: query, case detail display, aggregate view and advanced query.
Dataset: Scottsdale Health Center Chief Complaint
- Task 1: Describe the number of positive cases in the current dataset:
- Task 2: Find the time coverage (first and last case dates) in this dataset
- Date of first case: _____; Date of last case: _____
- Task 3: Identify the week with the highest number of cases.
- Task 4: Identify how many female patients with GI syndrome can be found within this dataset.
- Task 5: Identify the patient ID of the first patient with Botulism syndrome within the given dataset.
- Task 6: For the female patients in the age group 30-39, identify the top three syndromes besides "unknown":

Scenario 2: Spatial-Temporal Visualizer (STV)
This scenario, presented in two parts, will focus on the use of the STV tool to visually inspect the data distribution in both space and time. The user will be asked to identify information such as the peak number of cases, the area with the highest number of cases, and temporal and spatial distribution trends. The user will make use of the following tools provided in STV: Time Slider, GeoMap, Periodic Pattern Tool, Histogram and Timeline Tool.
Scenario 2-A
Dataset: User Study Test Dataset 1
- Task 1: Start STV with the User Study Test Dataset 1 and zip code boundary, isolate Botulism cases (by removing other syndromes from the map), and then identify the day of week with the most Botulism cases during the time period.
- Task 2: View the case distribution in Histogram tool and describe the temporal distribution.

- Task 3: Examine the spatial distribution using moving time window and expanding time window techniques and describe the spatial movement trend of Botulism cases.

Scenario 2-B

Dataset: Mesa Fire Department EMS data

- Task 1: Start STV with Mesa EMS data (between 9/1 2006 and 9/30 2006) and zip code boundary, change the color of categories with similar colors to avoid ambiguity, and then identify the hours of the most Trauma cases.
- Task 2: Identify the zip codes with the most Tox/Poison cases.
- Task 3: Identify the address of the youngest Cardio-Respiratory case in the Apache Junction Area (the most east zip code).
- Task 4: Isolate the Apache Junction area and identify the day of week with the most General Medicine cases:

Scenario 3: Hotspot Analysis

This scenario will target the use of the Hotspot Analysis tool embedded in the STV. The users will have 2 or 3 simulated datasets to evaluate. For each dataset, the user will be asked to identify outbreaks and investigate case details using STV.

Datasets: Under User Study Page (in our simulation process, the first 15-22 days are the made to be baseline data).

- Task 1: Regarding 167 Regarding Hemo syndrome cases with 1 injected outbreak, describe the outbreak you discovered.
- Task 2: Regarding 300 GI syndrome cases with 1 short term outbreak, describe the outbreak you discovered.

Public Health Affinity Domain: A Standards-Based Surveillance System Solution

Boaz Carmeli[1], Tzilla Eshel[1], Daniel Ford[2], Ohad Greenshpan[1], James Kaufman[2], Sarah Knoop[2], Roni Ram[1], and Sondra Renly[2]

[1] IBM Haifa Research Lab, University of Haifa, Mount Carmel, Haifa, 31905, Israel
{boazc,tzilla,ohadg,roni}@il.ibm.com
[2] IBM Almaden Research Lab, San Jose, CA, 95120, USA
{daford,kaufman,seknoop,srrenly}@us.ibm.com

Abstract. The negative impact of infectious disease on contemporary society has the potential to be considerably greater than in decades past due to the growing interdependence among nations of the world. In the absence of worldwide public health standards-based networks, the ability to monitor and respond quickly to such outbreaks is limited. In order to tackle such threats, IBM Haifa Research Lab and IBM Almaden Research Lab developed a PHAD implementation which consists of an information technology infrastructure for the public health community leveraging the Integrating the Healthcare Enterprise (IHE) initiative and important standards. This system enables sharing of data generated at clinical and public health institutions across proprietary systems and political borders. The ability to share public health data electronically paves the way for sophisticated and advanced analysis tools to visualize the population health, detect outbreaks, determine the effectiveness of policy, and perform forecast modeling.

1 Introduction

The 21st century healthcare environment is full of incompatible processes and technologies from proprietary systems. Increased population mobility and the effects of globalization necessitate the need to quickly and electronically share healthcare data within local communities as well as propagate significant data onward to regional, national, and international domains. In order to realize successful interoperability amongst all these systems, standardization is a recognized requirement.

Current standardization efforts are underway within both the clinical and public health sectors of our healthcare system. Unfortunately, these current efforts are largely independent of each other and this has resulted in divergent domain-specific standards. With little resources to implement even a single emerging standard, US vendors are now confronted with the following dichotomy to accommodate the clinical and public health workflows within their products.

Within the clinical domain, the Integrating the Healthcare Enterprise (IHE) initiative defines numerous technical frameworks for integrating many aspects of the clinical healthcare enterprise [1, 2]. The IHE Cross-Enterprise Document Sharing

D. Zeng et al. (Eds.): BioSurveillance 2007, LNCS 4506, pp. 147–158, 2007.

(XDS) profile is a technical specification which outlines a standards-based mechanism for sharing documents, including laboratory reports, among a group of healthcare organizations working together under a common set of policies and centralized services. This initiative has global collaboration and implementation.

Within the public health domain, the Centers for Disease Control (CDC) [3] is promoting Electronic Laboratory Reporting (ELR) [4] for transmission to the National Electronic Disease Surveillance System (NEDSS) [5] which is a component of the Public Health Information Network (PHIN) [6]. This initiative does not have global collaboration and implementation.

Public health institutions are significant consumers of clinical laboratory data as well as producers of laboratory reports performed by regional and national laboratories. By leveraging the same technical infrastructure and laboratory specific standards being created and actively adopted worldwide within the clinical domain, public health organizations can centralize laboratory data of interest like never before. This emerging standards-based clinical infrastructure has a wide array of public health partners and is an ongoing investment these partners are undertaking irrespective of public health interest. Taking advantage of this infrastructure will translate to improvement in reporting compliance, report completeness, and report accuracy due to the negligible cost overhead as a result of shared reporting standards. This is in stark contrast to the current policy of varying reporting requirements that necessitate unfunded custom implementations and hence result in poor compliance. Once laboratory data of interest is standardized and centralized, public health organizations can better focus attention on creating new tools to better visualize the population's health, detect outbreaks, determine the effectiveness of policy, and perform forecast modeling.

In the following section we summarize the worldwide IHE standards-based initiative with a focus on the Cross-Enterprise Document Sharing (XDS) specification. After explaining our chosen foundation, we describe the concept of a Public Health Affinity Domain (PHAD) and analyze our system's current capabilities. We then share what we have learned from a practical implementation of our first PHAD system. We conclude with a summary and our plans for future work.

2   Integrating the Healthcare Enterprise (IHE) Initiative

2.1   The IHE Organization

IHE is an international organization composed of healthcare and industry professionals whose aim is to improve the way computer systems in healthcare share information, and ultimately patient care, through the coordinated use of healthcare and IT standards. Each year since its inception, IHE produces a set of specifications, called *integration profiles*, which target a particular interoperability problem in a particular healthcare domain. These specifications are arranged in sets, called *technical frameworks*, which collectively target related interoperability problems in a healthcare domain. Within the technical framework, components which represent functional elements in the healthcare enterprise interact with each other as described in *use cases* in which *actors* participate in *transactions*. Actors are the conceptual representation of a physical system or systems in an enterprise. Each transaction is an exchange of messages between the participating actors. These technical frameworks are subjected to a yearly implementation and testing phase by real product systems. Additionally, upon test completion, the frameworks are evaluated and clarified as necessary. IHE efforts have been focused around clinical care, but new integration profiles relating to clinical trials, clinical research, and public health are emerging and are closely coordinated with those already established for clinical care.

2.2 XDS Integration Profile

One key profile is the Cross-Enterprise Document Sharing (XDS) integration profile depicted in Figure 1 [7]. XDS builds on industry standards, such as ebXML [8], SOAP [9] and HL7 v2 [10] to manage the sharing of clinical documents among healthcare enterprises. XDS assumes that participating organizations belong to one or more Clinical Affinity Domains. A Clinical Affinity Domain (CAD) is a group of Care Delivery Organizations (CDOs) that have agreed to work together using a common set of policies and share a common infrastructure. XDS establishes infrastructure for the CAD and interfaces for the CDO systems to enable sharing and exchange of medical information across CDOs.

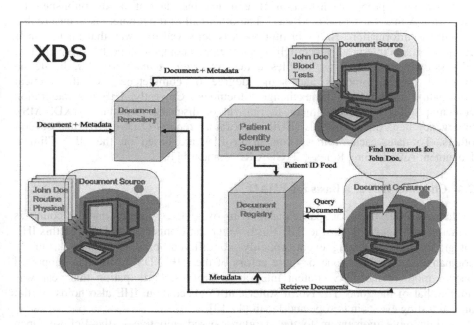

Fig. 1. The Cross-Enterprise Document Sharing (XDS) profile

We provide the following storyline to illustrate, in broad brush strokes, how XDS works.

John Doe is a patient at a small clinic with an IHE compliant Electronic Medical Record (EMR) system. His doctor electronically records John's vitals and a few other notes he has regarding the routine exam. This report and necessary metadata is automatically submitted by the EMR system, acting as a Document Source, to a

Document Repository in the CAD of which the small clinic is a member. The Document Repository stores the report, adds the URL to the metadata, and forwards it all onto the Document Registry for the CAD.

John's doctor supports John's decision to have an HIV screening test. John goes to an affiliated hospital's laboratory at the other end of town which is also a member of the CAD. John has his blood taken and the lab work is processed a few days later. This lab report is also submitted to the Document Registry of the CAD.

John's doctor receives notification that the lab work has been completed, so he goes to his EMR system, which now acts as a Document Consumer, to obtain the physical exam report and the lab report for review. The EMR system executes a query against the CAD Document Registry to find all the clinical documents for John Doe. The Document Registry is able to answer the query and return a set of links for John's clinical documents to the EMR system which then presents them, as appropriate, to John's doctor.

2.3 XDS Integration Content Profiles

The IHE XDS integration profile only solves half of the clinical document exchange interoperability problem; institution B can get the data it needs published by institution A in a standards-based way. The other half of the problem is to ensure that the clinical information sent by institution A is conveyed in a way that institution B can process and interpret. Hence, IHE has developed companion profiles to XDS that address this very issue. These profiles, called *integration content profiles*, dictate the expected format, concepts and some degree of completeness of the clinical information contained in the exchanged document. Currently, IHE has integration content profiles for general medical, referral or discharge summaries (XDS-MS), emergency department referrals (EDR), laboratory reports (XDS-LAB), and several others. To date, each of these content profiles are based on the HL7 Clinical Document Architecture, Release 2 (CDA R2) standard [10].

2.4 Choosing IHE as a Basis for PHAD

In January 2007 over 350 participants from over 70 companies and 12 countries gathered in Chicago, IL for the IHE North American Connectathon [11]. In this IHE integration profile testing event, roughly 40 different product systems from 25 companies implemented one or more actors of the IHE XDS profile and supported one or more of the XDS content integration profiles. Participation this year was double that of the 2006 IHE North American Connectathon. IHE also holds similar annual testing events in Europe and Japan [12,13].

Adding more momentum to the standards-based initiative is the Eclipse Open Healthcare Framework (OHF) [14, 15]. Eclipse is an open source community whose projects are focused on providing an extensible development platform and application frameworks for building software [16]. OHF is a project within Eclipse formed for the purpose of expediting healthcare informatics technology with contributing companies including Mayo Clinic [17], Jiva Medical [18], Inpriva [19], and IBM [20]. OHF currently provides implementations of the XDS Document Source and XDS Document Consumer as well as implementations of several actors from other integration profiles that are necessary compliments to the XDS profile.

Using worldwide adopted standards and sharing existing infrastructure with the clinical domains, public health has much to gain from this uninterrupted flow of data across previous barriers.

3   A Public Health Affinity Domain

3.1   The PHAD Concept

A Public Health Affinity Domain (PHAD) is a concept we adopted that is similar to the Clinical Affinity Domain (CAD). A PHAD has a number of members, including the Care Delivery Organizations (CDO), which may be active participants in existing CADs. In addition to clinical data sources, a PHAD can have contributors that report on veterinary, food, environment, and workplace surveillance activities.

Public health data gathering today is hierarchical in nature. In the US, laboratory results of interest are propagated upwards from local entities, to the county and state level, and to the national CDC level as necessary. A PHAD improves upon this model by formalizing the report itself, centralizing the data at a shared governance point, and enabling cross-domain document referencing as a report is automatically escalated from one level to another.

The PHAD hierarchy described below in Figure 2 is similar to the current US pathway for public health information exchange with the exception that international collaboration is not only technically feasible but encouraged. The model consists of several regional PHADs feeding into a national PHAD and several national PHADs feeding into an international PHAD.

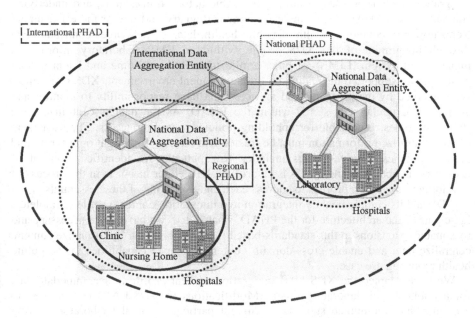

Fig. 2. Sample hierarchical PHAD arrangement for two countries

The *regional PHAD* is the union of an invested community of public and private organizations working in conjunction with public health officials. In the US, this PHAD is consistent with the local, county, and state agencies. Common laboratory-based surveillance vary by state but often include the incidence of newborn screening disorders, food-borne diseases, sexually transmitted diseases, virus typing, as well as water, food, and environmental quality.

The *national PHAD* is a centralized government organization whose responsibility it is to aggregate data from the regional PHADs and perform surveillance across the regional PHAD domains. The Centers for Disease Control (CDC) is consistent with this concept. The CDC produces aggregate reports for national government agencies as well as bodies such as the World Health Organization (WHO). When the country has one or more international surveillance partnerships, the national PHAD feeds significant data forward to the international PHAD.

The *international PHAD* is an open collaboration between a set of countries recognizing that infectious diseases do not recognize political or geographic boundaries. Several partnerships are forming that are consistent with this concept, such as within the European Union and the Middle East. These partnerships have a defined (limited) scope for sharing reports and strong citizen privacy policies. The international PHAD receives data from partner national PHADs.

While we haven't yet proposed a global PHAD, our technical infrastructure is flexible and easily extended to yet another hierarchical layer, such as the WHO.

3.2 The PHAD Architecture

The architecture for the PHAD builds directly upon XDS and its integration and integration content profiles: it utilizes the same actors, transactions, and underlying standards. Each PHAD has a single Document Registry and one or more Document Repositories to contain a wealth of public health data. XDS Sources contribute data through integration of the XDS profile within an EMR, laboratory information management system (LIMS), or other application. Both real-time and latent upload strategies are supported depending on the deployment environment. XDS Consumers are embodied by the programs and agencies that have responsibility to monitor and aggregate particular data stored within the PHAD. As the data is centralized, the PHAD becomes responsible for upholding policies regarding access. The ownership of any centralized information must be trusted and have well-thought out sharing and privacy policies. These policies include, how patients are identified, consent, if necessary, is obtained, and access is controlled. IHE either has or is in the process of developing integration profiles to address each and every one of these essentials.

XDS and its complementary integration and integration content profiles complete a large part of the architecture for the PHAD. That stated, we have made necessary and reasonable extensions to this standards-based infrastructure to truly formalize content, centralize data and enable cross-domain content reference capabilities in the public health reporting use case.

We have adapted the XDS-LAB integration content profile to accommodate data for a public health laboratory report. Modifications to XDS-LAB were needed to accommodate non-human subjects, document participants in the laboratory testing process, and to group tests for a reportable condition in a consistent manner. A benefit to adopting XDS-LAB is that the communication loop is closed between clinical care and public health reporting. Public health receives the initial clinical XDS-LAB report, for example "Identified Salmonella species", and can return the final epidemiology XDS-LAB report back to the clinical care provider for inclusion in the patient's medical record, for example "Identified Salmonella serotype virchow".

We enhanced the query capabilities of the Document Consumer edge systems beyond the minimal query set defined by the XDS profile. We accomplished this by relaxing query parameters and by extending the Document Registry metadata to include coded values for reportable condition, specimen type, and other public health concepts. Finally, we enabled document relationships already existing within the context of XDS Document Registry to span across different PHADs.

We have focused on laboratory-based surveillance. The infrastructure can be expanded to include additional standards-based data sharing going forward.

4 A Practical PHAD Implementation

PHAD principles are illustrated by example in the solution we created in collaboration with participating countries as part of their shared mission of addressing emerging infectious disease and strengthening international public health efforts. Their goals are to build an infrastructure within each nation for disease surveillance and outbreak response as well as foster cross-border communication.

We present here our adaptation of the layered architectural PHAD concept to the diverse organizational structures and workflows that span across the participating countries. A high level overview is shown in Figure 3.

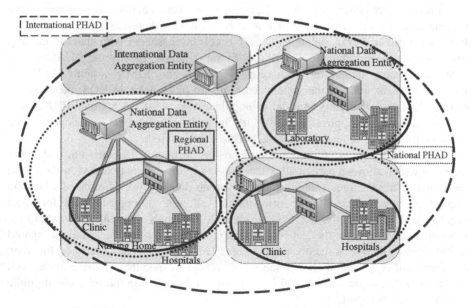

Fig. 3. High level overview of the implemented PHAD architecture

The regional PHAD is composed of central laboratories (similar to a state public health laboratory), sentinel physicians, and clinical laboratories. In this case, the two countries have chosen to share a regional PHAD. The national PHAD is to be maintained within each country's CDC. These PHADs are solely responsible for forwarding documents onto the international PHAD that meet the policies agreed upon by all the member countries.

There are several complications. Certain designated sentinel physicians and clinical laboratories report some data directly to the CDC rather than and sometimes in addition to the central laboratory. Policies supporting each distinct requirement need to be accommodated. Furthermore, this multi-path convergence of data flow complicates the task of duplicate submission identification. Such a situation is possible when a local entity reports to both the CDC and the central laboratory (red connector above) while the regional PHAD forwards additional information to the CDC as a second report without specific knowledge that the CDC has a preliminary report.

As the document is propagated from the regional to the international PHAD, less data is going to be shared. Original data sources can be clinical documents. These, being the most complete, will contain full patient identification, full authorship and participant information, and all the test results that were ordered together. Reports to the national PHAD will receive limited patient identification details, partial authorship information, no participant information, and only the reportable test result. The international PHAD will receive only broad spatial identification of the incidence location, general purpose authorship, no participant information, and only the final summary result.

Current aggregation and surveillance tools vary across the participating organizations in this hierarchy. Much as it is in the US, there are still a lot of paper log books, paper forms and reports, and excel spreadsheets with macros holding together the infrastructure for public health surveillance. Ideally the PHAD would leverage existing underlying technology – EMR and LIMS systems deployed. Unfortunately, as none of the systems currently in place have an XDS interface, we rely on duplicate data entry in order to populate reports for the regional PHAD. We have provided a web application, shown in Figure 4, by building atop the open source software for XDS provided through the Eclipse OHF project.

The web application presents a simple report formation, submission, and query interface which allows a participating laboratory to publish data to and view data from the shared infrastructure. Tedious and error-prone data entry is eliminated when the clinical partner is part of the infrastructure allowing a copy-forward for most requisition data. Currently, this feature is enabled through a patient query where the laboratory user obtains the patient document of interest by querying the shared infrastructure and selects the document to use. This feature is particularly useful when the document is the initial clinical laboratory report as specimen information as well as the globally unique document id is available for inclusion in the subsequent public health laboratory report.

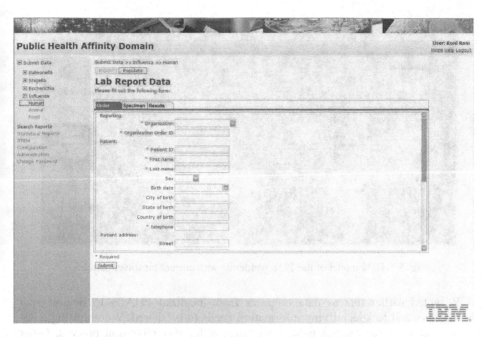

Fig. 4. Screen shot of web form for ordering and reporting

Also, as a beginning to new surveillance tooling, we also incorporated a project soon to be released in OHF, the Spatiotemporal Epidemiological Modeler (STEM) framework [21]. This tooling framework has a built in geographical information system (GIS) for dynamically running current conditions as well as model predictions. It will be released with a significant number of starter maps which can be enhanced and redistributed as experts begin to add their domain knowledge to the STEM community. Figure 5 shows a current model of the 1918 pandemic flu given contemporary population density and air travel pattern data. Once a model of infection is derived, multiple simulations can be started to examine scenarios based on actions taken, such as closing airports and schools, or vaccinating/culling vectors of the infection. This framework has been integrated within our PHAD through the query aggregation of pertinent data in the PHAD system. Data is aggregated by time and geographic location. The model can be run in regular time increments or by selecting a time-frame for display.

Finally, we are in the process of integrating reporting tools and de-identification technology. These extremely valuable and necessary features present some interesting challenges. As a basis for our report generation capabilities, we are relying on an open source project from Eclipse called Business Intelligence and Reporting Tools (BIRT). BIRT creates charts, graphs, and reports using SQL or XPATH. Our users are most familiar with basic SQL while our infrastructure is XML based, and many XML documents may be retrieved in a single query for which XPATH was not designed. Our goal is to hide the underlying complexity while giving our users the easiest report writing experience.

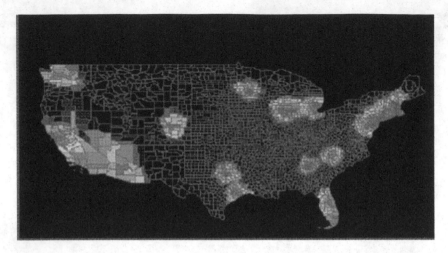

Fig. 5. STEM model of the 1918 pandemic with current air travel patterns

We noted earlier that as data migrates from localized PHADs to international PHADs less and less identifying information needs to be shared. We are using a de-identification tool to encrypt patient information for this migration process. Other options include the complete removal of such data, but this will present a significant challenge when it comes to identification of duplicate reports for the same patient. This is an issue when one report author submits to more than one PHAD, such as we saw with a clinical participant submitting to both the regional and national PHADs. Re-identification is also important when public health officials identify an outbreak and require more information from each case.

5 Summary

Global public health in the 21st century faces many challenges. Information technology can enable collaboration across political boundaries supporting different social contracts and governmental organization structures. While creating interoperable systems requires political will and close collaboration, interoperability work of the sort done to support the development of electronic health records for individual patients and clinicians provides a base for efforts in the public health domain. By leveraging the same technical infrastructure and laboratory specific standards being created and actively adopted worldwide within the clinical domain, public health organizations can centralize laboratory data of interest like never before. Once this is accomplished, public health organizations can better focus attention on creating new tools to better visualize the population's health, detect outbreaks, determine the effectiveness of policy, and perform forecast modeling.

As previously noted, XDS and its complementary integration and integration content profiles complete a large part of the architecture for the PHAD, however during our implementation, extensions to this standards-based infrastructure to truly formalize content, centralize data and enable cross-domain content reference capabilities in the public health reporting use case became apparent.

A vital key to widespread adoption of XDS across both clinical and public health domains is work being done through open source initiatives like the Eclipse OHF project to greatly ease the adoption of healthcare IT standards. These initiatives give vendors the ability to focus on the competitiveness of their product rather than trying to keep up with complicated and evolving standards.

6 Future Work

In this work we laid the groundwork for additional research and collaboration with both clinical and public health partners to pursue a more unified path towards standardization across all of healthcare, ultimately leading to improvements in early detection and possible prevention of a worldwide disaster due to infectious disease. We have learned much from our initial implementation of XDS in a public health environment and are in the process of feeding our experiences back into IHE and the relevant standards bodies with the ultimate goal of growing a Public Health sub-domain in the IHE along with members from PHDSC [22], HITSP [23] and HL7. We will begin with the collaborative development of an integration content profile for Public Health Case Report for Laboratory Data, based on our adaptation of XDS-LAB this year. This will enable us to do a more complete and collaborative evaluation of HL7 CDA R2 as a medium for public health reporting. This work and the future standards-based profiles for public health will require testing by the international community and the hard, ongoing work of early adopters.

The work that will be done in building the Public Health sub-domain within IHE will help to expand the PHAD solution to more types of public health surveillance systems. We conclude that while more work is needed we believe our approach offers many benefits to the public health domain and can significantly help to gather data required for infectious disease surveillance.

References

1. IHE: Integrating the Healthcare Enterprise - http://www.ihe.net/
2. J. H. Kaufman, I. Eiron, G. Deen, D.A. Ford, E.Smith, S.Knoop, H. Nelkin, T. Kol, Y. Mesika, K. Witting, K. Julier, C. Bennett, Bill Rapp, "From Regional to National Health Information Infrastructure", Perspectives in Health Information Management, 2;10 Fall 2005, AHIMA
3. CDC: Centers for Disease Control - http://www.cdc.gov/
4. ELR: Electronic Laboratory Reporting - http://www.cdc.gov/nedss/ELR/index.html
5. NEDSS: National Electronic Disease Surveillance System - http://www.cdc.gov/nedss/
6. PHIN: Public Health Information Network - http://www.cdc.gov/phin/
7. XDS: Cross-Enterprise Document Sharing - http://www.ihe.net/Technical_Framework/index.cfm#IT
8. ebXML: Electronic Business using eXtensible Markup Language - http://www.ebxml.org/
9. SOAP: Simple Object Access Protocol - http://www.w3.org/TR/soap/
10. HL7: Health Level Seven - www.hl7.org

11. IHE Connectathon - http://www.ihe.net/Connectathon/index.cfm
12. IHE Europe - http://www.ihe-europe.org/
13. IHE Japan - http://www.ihe-j.org/en/
14. Eclipse OHF: Eclipse Open Healthcare Framework - http://www.eclipse.org/ohf
15. Eclipse - http://www.eclipse.org/
16. Srivatsava R. Ganta, Eishay Smith, Sarah E. Knoop, Sondra R. Renly, James H. Kaufman. The Eclipse Open Health Framework, HCTM 2006; 5th International Conference on the Management of Healthcare & Medical Technology, 2006.
17. Mayo Clinic - http://www.mayo.edu/
18. Jiva Medical - http:// www.jivamedical.com/
19. Inpriva - http:// www.inpriva.com/
20. IBM - http://www.ibm.com/
21. STEM: Spatiotemporal Epidemiological Modeler - http://wiki.eclipse.org/index.php/STEM/
22. PHDSC: Public Health Data Standards Consortium - http://phdatastandards.info/
23. HITSP: Healthcare Information Technology Standards Panel - http:// www.ansi.org/hitsp/

The Influenza Data Summary: A Prototype Application for Visualizing National Influenza Activity

Michelle N. Podgornik[1], Alicia Postema[2], Roseanne English[1], Kristin B. Uhde[1], Steve Bloom[1], Peter Hicks[1], Paul McMurray[1], John Copeland[1], Lynnette Brammer[2], William W. Thompson[2], Joseph S. Bresee[2], and Jerome I. Tokars[1]

[1] Centers for Disease Control and Prevention, National Center for Public Health Informatics, Division of Emergency Preparedness and Response, Atlanta, GA 30333, USA
[2] Centers for Disease Control and Prevention, National Center for Immunization and Respiratory Diseases, Influenza Division, Atlanta, GA 30333, USA
{MPodgornik,APostema,RXE1,KUhde,SBloom1,PHicks,PMcMurray,
JCopeland,LBrammer,WThompson1,JBresee,JTokars}@cdc.gov

Abstract. The Influenza Data Summary (IDS) is a tool that provides a unified view of influenza activity in the United States. It currently incorporates data from portions of the U.S. Influenza Surveillance System and BioSense. The IDS allows users to customize dashboards, interactive maps, and graphs from each of these data sources. The purpose of this paper is to provide an overview of the IDS and to discuss current features and future plans for improvement.

Keywords: syndromic surveillance, influenza.

1 Introduction

Influenza is an acute respiratory illness caused by infection with influenza type A or influenza type B virus. Characteristic symptoms of uncomplicated influenza illness include abrupt onset of fever, cough, sore throat, headache, extreme fatigue, and muscle aches. Influenza can also cause severe illness and death due to pneumonia (primary viral or secondary bacterial pneumonia) or by exacerbating underlying medical conditions [1, 2, 3]. Both influenza A and influenza B viruses are constantly changing through a slow process called antigenic drift resulting in annual epidemics of disease. During annual influenza epidemics in the U.S., between 5%-20% of the general population may be infected, and an average of approximately 200,000 persons will be hospitalized, and approximately 36,000 persons will die [4, 5, 6]. Influenza A viruses can also change more dramatically resulting in a new influenza A subtype that, if transmissible from person to person, could cause a pandemic. For planning purposes, the U.S. Department of Health and Human Services Pandemic Influenza Plan assumes that the clinical attack rate of the next influenza pandemic will be 30% in the overall population [7]. Further, it is estimated that a moderate influenza pandemic might result in 865,000 hospitalizations and 209,000 deaths, while a severe pandemic might lead to 9.9 million hospitalizations and 1.9 million deaths [7].

D. Zeng et al. (Eds.): BioSurveillance 2007, LNCS 4506, pp. 159–168, 2007.

Annual vaccination with influenza virus vaccine is the best way to mitigate the impact of influenza. Antiviral drugs used for treatment or prophylaxis are adjuncts to the vaccine that can also help reduce the impact of influenza. Good personal hygiene habits such as covering one's nose and mouth when sneezing and coughing and frequent handwashing can also help reduce the spread of disease.

2 Influenza Data Summary

Surveillance for influenza in the United States is accomplished primarily through the U.S. Influenza Surveillance System, a multi-component viral and disease surveillance system that is coordinated and managed by the Influenza Division at the Centers for Disease Control and Prevention (CDC) and involves state health departments, state public health laboratories, vital registrars offices, and clinicians across the country. In recent years, the increasing availability of both electronic health data and biosurveillance systems such as the CDC BioSense project have made it possible to explore the utility of using these types of data to track influenza activity. The Influenza Data Summary (IDS) was created as a tool to display influenza data from disparate sources including traditional influenza surveillance systems and BioSense and present them to decision makers in a common user interface and format.

The IDS will improve the ability of public health officials at all levels to monitor influenza activity across the nation and to provide health situational awareness of both seasonal and non-seasonal influenza. Potential benefits include better visualization of local, state, and national patterns of influenza activity, timely and easy to understand surveillance data, and standardized formats for data obtained from different sources. These capabilities will allow for faster comparisons of data across seasons and geographic regions and assist with informing policy and guiding public health control measures.

The IDS currently contains data from three of the seven U.S. Influenza Surveillance System components and three types of data from BioSense. Each currently displayed data source is described below and efforts are underway to incorporate additional data sources.

2.1 U.S. Influenza Surveillance System

The U.S. Influenza Surveillance System is comprised of seven components: laboratory surveillance, outpatient influenza-like illness (ILI) surveillance, pneumonia and influenza related mortality surveillance, pediatric mortality surveillance, assessment of influenza activity at the state level, and hospitalization surveillance through the Emerging Infections Program (EIP) and the New Vaccine Surveillance Network (NVSN). Data are reported weekly from October through mid-May for all system components. Since 2001, a growing subset of laboratories and health care providers report influenza data to CDC year round. Mortality data from the 122 Cities Mortality Reporting System were already available year round and the pediatric mortality reporting system has been a year round system since it began in the 2004-05 season.

All data reported to CDC as part of the national influenza surveillance system are made available to state health departments in real-time or close to real-time depending on the reporting mechanism used. State health department officials have access to outpatient ILI and laboratory data in real time on a password protected website and can use these data to inform local decisions and follow-up on any unusual events. City specific 122 Cities Mortality data and state specific pediatric mortality data are posted weekly in the Morbidity and Mortality Weekly Report (MMWR) and summary data for these systems, EIP, NVSN, and the state and territorial epidemiologists' report are available from the weekly influenza surveillance report produced by CDC's Influenza Division from October through mid-May. The report includes national and regional level data and is widely disseminated nationally and internationally, and available to the general public on the Internet. The IDS currently receives data from three of these sources:

- Influenza virus data are obtained weekly from approximately 85 U.S. World Health Organization (WHO) and 65 National Respiratory and Enteric Virus Surveillance System (NREVSS) Collaborating Laboratories that report the number of specimens tested for influenza and the number positive by influenza type (A or B) and in some cases subtype (A/H1 or A/H3).
- Outpatient ILI data are obtained weekly through the U.S. Influenza Sentinel Provider Surveillance Network, a network of approximately 2,400 primary care providers representing all 50 states. Providers report the number of outpatients seen for any reason and the number of those with ILI by age group.
- The State and Territorial Epidemiologists Reports provide overall levels of activity within each state, the District of Columbia, New York City, and Puerto Rico. Influenza activity is described as either no activity or sporadic, local, regional, or widespread activity.

2.2 BioSense

BioSense is the nation's real-time electronic biosurveillance system. BioSense receives health data in electronic format from several types of sources, analyzes these data, and presents them through a secure Internet-based application. The BioSense application is available to qualified hospital and public health users via the CDC Secure Data Network. The system enables both health situational awareness and early event detection by identifying and confirming possible events and tracking and managing their size and spread. Currently, BioSense contributes the following data to the IDS:

- Outpatient facilities associated with the Department of Defense (DoD) (n>300) and Department of Veterans Affairs (VA) (n>800). Demographic, diagnostic, and procedure data are received daily from these sources. Data included in the IDS are weekly summaries of outpatient visits with an ICD-9-CM diagnosis code of 487 (i.e., the code for influenza).

- BioSense Hospitals. Real-time data is received from over 350 hospitals. Most hospitals send demographic, chief complaint, and diagnostic data for emergency department (ED) patients, and many send these data for hospital inpatients and outpatients as well. Data currently used in the IDS includes weekly summaries of chief complaint and physician diagnosis data on inpatients, outpatients, and emergency department patients. For the chief complaint data, records containing the words "influenza" and "flu" are assigned to the ILI category. As with the VA and DoD data, physician diagnoses of ICD-9-CM code 487 are tracked.

Influenza test results from commercial laboratories may be included in future versions of the IDS.

3 Methods

The IDS is a prototype application built using SAS Business Intelligence* (SAS BI) software to gather, provide access to, and display data in a common environment. The IDS tool improves the timeliness and quality of data visualization and helps in the analysis of trends. The prototype application is currently available only to CDC users. However, there are plans to demonstrate the IDS to state and local users outside CDC so that their comments and suggestions can be incorporated during the ongoing development of this tool. An application that can be accessed by users outside CDC, including public health personnel at all levels as well as applicable healthcare personnel, is scheduled for the beginning of the 2007-2008 influenza season.

*Use of trade names is for identification purposes only and does not imply endorsement of SAS Institute or its subsidiaries by the Centers for Disease Control and Prevention.

4 Data Analysis and Presentation

The current screens on the IDS include the following:

- **Flu Kiosk.** This page has 1) a brief influenza status report, which contains bullets or other text summary of current national influenza activity as described in the weekly influenza surveillance report, 2) links to influenza-related websites and news feeds, and 3) a dashboard (see Fig. 1). The content of this screen is set by CDC personnel. The dashboard has approximately 16 thumbnail screenshots of national maps and time series graphs that are automatically refreshed with the latest available data. These dashboard views were created by saving views from the Time Series or Map screens (see below). By clicking on a button, the user can go to a full-screen view of the dashboard content.

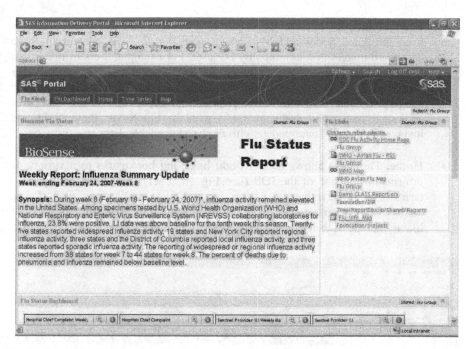

Fig. 1. Screenshot of the Flu Kiosk that shows the influenza status report and the links to related websites and news feeds

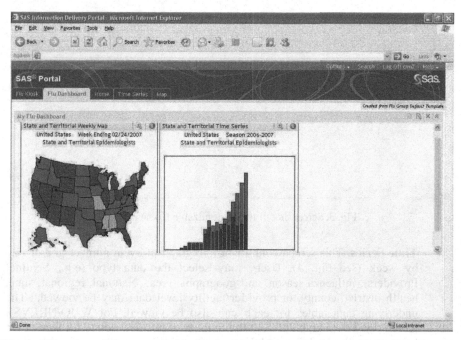

Fig. 2. Screenshot of the customizable Flu Dashboard that shows an example of a map and a time series graph

- **Flu Dashboard.** This screen consists of a variable number of saved images from the Time Series and Map pages (see Fig. 2). The content of these dashboard images is determined by the user. For example, users from a given geographic area can create a summary similar to the Flu Kiosk, but specific for their area. Thus, the user can simultaneously view all data applicable to their geographic area. The user can rename this screen and create and save additional customized dashboards. As with the Flu Kiosk, these dashboard images can be set to auto-update with the latest available data.

- **Home.** This page allows users to add links and bookmarks to a variety of information outside of the IDS (see Fig. 3). Users can search for specific terms (e.g., avian influenza) within external files, applications, news feeds, and websites. Users are then able to choose which of the results they would like to include in their Home page.

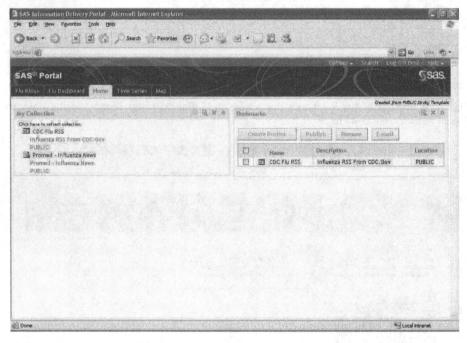

Fig. 3. Screenshot of the customizable Home page

- **Time Series.** This screen contains a line graph of event number and/or rate by week (see Fig. 4). Users may select the data type (e.g., Sentinel Providers), influenza season, and geographic area. National, regional, state, health district, county, or provider/facility level data may be viewed. The underlying data tables for each can also be viewed. For WHO/NREVSS laboratory data, a graph of both the number of positive isolates broken down by virus type and subtype and the percent positive by week in the selected flu season is also displayed.

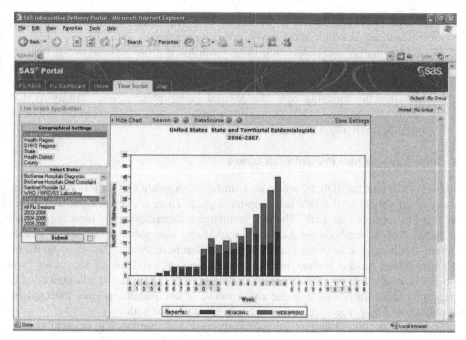

Fig. 4. Screenshot showing a full-screen view of the time series graph from Figure 2. Modifications available to users are displayed on the left-hand side of the page.

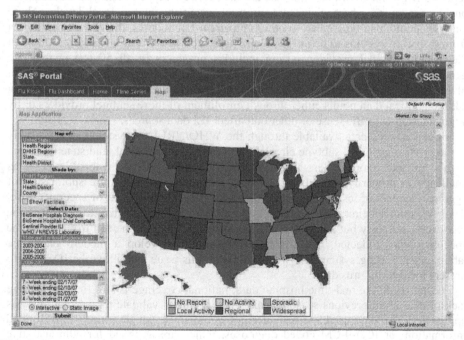

Fig. 5. Screenshot showing a full-screen view of the map from Figure 2. Modifications available to users are displayed on the left-hand side of the page.

- **Map.** As with the Time Series screen, the user can select data type, influenza season and week, and geographic area (see Fig. 5). Additionally, users can click on geographic regions to drill down to finer geographic detail. For outpatient ILI data, the number of facilities (or providers), ILI visit count, total visit count, and ILI rate can be easily obtained by mousing over the desired geographic region. Analyses that show the change in the event from a baseline period, or from the previous month, are being developed.

4.1 Data Analysis and Presentation Issues

Combining data in the IDS to arrive at a unified, comprehensive view of influenza activity at all geopolitical levels is a long-term goal. There are at least two significant challenges to meeting this goal. The first challenge is determining the most appropriate and useful analytic methods for each data source and also across sources that include different data elements that were collected using different methods. The other significant challenge relates to the limited number of data points available at the more local geopolitical levels. For some systems included in the IDS, this is due to a limited number of reporting sites; for others, it is due either to use of antiquated reporting mechanisms that only allow for aggregate reporting or to a combination of both.

For the Sentinel Provider data, comparing geographic-specific (e.g., state to state or even provider to provider) ILI rates may not be valid because individual providers have different baseline ILI rates. Instead, analyses that compare each provider's reported ILI to their own baseline rate are needed. This type of analysis would allow one to determine where elevated activity is occurring at the local level. These analyses could also be applied to other outpatient ILI sources, thereby allowing data from disparate systems to be analyzed and presented in a uniform manner. Although the current sentinel provider system includes providers in all 50 states, they are generally found in the population centers of the state. The ability to track the impact of influenza on outpatient visits at the local level would be improved by including additional local data points either through enrollment of more sentinel providers or by augmentation with BioSense or other outpatient ILI data sources.

Influenza virus data available through the WHO/NREVSS system can adequately describe virus type and subtype circulation at a national, regional, and state level but, with a few exceptions, is less able to provide this information at a local level. This challenge is primarily a result of reporting mechanisms that require submission of aggregate counts of test results rather than electronic transmission of specimen level data. Without specimen level data, it is possible to link the positive test result to the reporting laboratory but not to the location of residence of the person from whom the specimen was collected. Efforts are underway to develop a modern electronic laboratory reporting system that would alleviate this issue and allow the IDS tool to present local level virus data.

For the BioSense real-time hospital data, national coverage is still limited, and the lack of data from previous seasons makes it difficult to validate the current analysis approach. The data used are patient chief complaints, which convey only limited information, or ICD-9-CM coded diagnoses, which are assigned for administrative purposes, rather than cases meeting a formal definition. The ongoing recruitment of hospitals that provide real-time data to BioSense will improve the current issue with geographic coverage. Multiple patient classes (i.e., inpatient, outpatient, and ED) and data types (e.g., chief complaint, working diagnoses, and final diagnoses) have different latencies and definitions, making comparison difficult.

BioSense DoD/VA data also represent ICD-9-CM codes assigned for administrative purposes rather than for cases meeting a formal definition. Although there is broad geographic coverage of BioSense DoD and VA facilities, these data are not demographically representative of the United States population because coverage is limited to populations served by military or Veterans Affairs facilities. However, the data do correlate well with Sentinel Providers data nationally and in many high-volume states [8].

5 Findings

Comparisons of data obtained from several different data sources in the current influenza season exemplify some of these analysis issues. As of the date of preparation of this paper (January 24, 2007), the following patterns have been observed. Nationally, surveillance data showed increasing levels of influenza activity during October 2006 – December 2006 and a slight decline in activity in January 2007. The decline noted in January could be due in part to the fact that some areas of the country experienced activity early in the season that began to subside just as activity in other parts of the country began to increase. It could also be due in part to an artificial elevation of reported activity during the holidays. The holiday artifact occurs because fewer people seek medical care for preventive reasons. Therefore, the larger proportion of persons seeking medical care for ILI is a result of a smaller denominator (number seeking care for other reasons) rather than a larger numerator (number of ILI cases).

Of note, the Sentinel Providers ILI visit rate, the WHO/NREVSS percent of isolates testing positive, and the BioSense hospital chief complaint ILI rate all showed a similar picture of increasing and decreasing activity nationally during October 2006 – January 2007. The comparison of these trends in varying data sources was facilitated by the IDS application, specifically the dashboard views which allow side-by-side comparisons. BioSense hospital diagnosis and BioSense DoD/VA diagnosis data did not track as closely with the other data types. Reasons for this discrepancy may include differing patient populations covered and the need to include additional diagnosis codes (i.e., those for pneumonia) when analyzing certain diagnosis data. The IDS also facilitates comparison of multiple data types within individual states and other geographic units, but this information is too complex to summarize in this current report.

6 Plans

Future plans for the IDS include:

- Engaging public health partners in continued development of the system as well as developing methods for jurisdiction-specific access so that the system can be available to users outside CDC.

- Continued improvement of data analysis methods, including the use of weights to adjust for differing population coverage in different states, stratified analysis to show rates specific for age and other factors, and trend analyses such as the EARS CuSUM method for comparisons of reported ILI to a provider-specific baseline.
- Incorporating additional data sources into the IDS. Such sources may include the 122 Cities Mortality Reporting System and the Countermeasure and Response Administration system, which tracks the distribution of vaccines, antibiotics, and antivirals in response to public health outbreaks or preparedness measures.

7 Conclusion

Creation of the IDS was prompted by concern about pandemic influenza and the recent availability of additional potential data sources for monitoring influenza. Using programming and analysis tools in a commercially-available software package, we rapidly created an application that displays multiple data types in a user-friendly, customizable format. The prototype has been used at CDC during the current influenza season but requires additional input and development to maximize its utility and to ready it for wider release. In addition to its use for influenza, the IDS may serve as a prototype for collating and co-displaying a variety of data for other diseases of public health importance.

References

1. Heymann, D.L. (ed.): Control of Communicable Diseases Manual. 18th edn. American Public Health Association (2004)
2. Nicholson, K.G.: Clinical features of influenza. Semin. Respir. Infect. 7 (1992) 26-37
3. Douglas, R. Jr.: Influenza in man. In: Kilbourne, E.D. (ed.): Influenza Viruses and Influenza. Academic Press, Inc., New York, NY (1975) 395-418
4. Monto, A.S., Kioumehr, F.: The Tecumseh Study of Respiratory Illness. IX. Occurrence of influenza in the community, 1966-1971. Am. J. Epidemiol. 102 (1975) 553-563
5. Thompson, W.W., Shay, D.K., Weintraub, E., Brammer, L., Bridges, C.B., Cox, N.J., Fukuda, K.: Influenza-associated hospitalizations in the United States. JAMA. 292 (2004) 1333-1340
6. Thompson, W.W., Shay, D.K., Weintraub, E., Brammer, L., Cox, N., Anderson, L.J., Fukuda, K.: Mortality associated with influenza and respiratory syncytial virus in the United States. [see comment]. JAMA. 289 (2003) 179-186
7. Department of Health and Human Services.: HHS Pandemic Influenza Plan: Part 1 – HHS Strategic Plan (available at http://www.hhs.gov/pandemicflu/plan/. Accessed 01/29/2007)
8. Tokars, J., Roselle, G., Brammer, L., Pavlin, J., English, R., Kralovic, S., Gould, P., Postema, A., Marsden-Haug, N.: Monitoring influenza activity using the BioSense System. Adv. Dis. Surveill. 1 (2006) 70

Global Foot-and-Mouth Disease Surveillance Using BioPortal

Mark Thurmond[1], Andrés Perez[1], Chunju Tseng[2], Hsinchun Chen[2],
and Daniel Zeng[2]

[1] Foot-and-Mouth Laboratory, University of California, Davis, CA U.S.A
[2] Artificial Intelligence Laboratory, University of Arizona, Tuscon, AZ U.S.A
mcthurmond@ucdavis.edu, amperez@ucdavis.edu, chunju@u.arizona.edu,
hchen@eller.arizona.edu, zeng@email.arizona.edu

Abstract. The paper presents a description of the FMD BioPortal
biosurveillance system (http://fmd.ucdavis.edu/bioportal/) that is
currently operating to capture, analyze, and disseminate global data on
foot-and-mouth (FMD) disease. The FMD BioPortal makes available
to users world-wide FMD-related data from the Institute for Animal
Health at Pirbright, England. The system's tools include those for tabu-
lating and graphing data, performing spatio-temporal cluster analysis of
outbreak cases of FMD, and analyzing genomic changes in FMD viruses.
The FMD BioPortal also includes the FMD News (http://fmd.ucdavis.
edu/index.php?id=1), which is a near real time web search to identify and capture FMD-related news items appearing worldwide. Major systems components include a communication backbone for secure,
real-time data transfer, a data analysis module that can run analytical programs to assess spatial-temporal clustering, and an interactive
visualization tool for integrated analysis and display of epidemiological
and genomic data.

Keywords: Foot-and-mouth disease, BioPortal, surveillance.

1 Introduction

Foot-and-mouth disease (FMD) is a highly contagious disease that imposes severe and far-reaching economic consequences for countries that acquire the disease. The disease is caused by the FMD virus (FMDV), which is considered to
be one of the most infectious disease agents known for cloven-hoofed mammals.
Major economic and social impacts of the disease occur world wide as a result
of inefficient animal production and restrictions on trade, which can manifest in
limiting national development and socio-economic growth. The disease is spread
by transmission of exhaled or excreted virus through direct physical contact between animals, such as cattle, pigs, sheep, and goats, and by indirect contact
with fomites containing infectious virus, such as contaminated vehicles, feed, or
clothing of livestock personnel. As a consequence, increased contact among animals, such as would occur at markets or following transportation of animals to

D. Zeng et al. (Eds.): BioSurveillance 2007, LNCS 4506, pp. 169–179, 2007.

new areas, increases the rate of transmission. Disease spread in endemic regions of the world tends to follow animal and human movement patterns, where the rate of new outbreaks can be expected to increase at times of increased transportation or movement of animals or animal products. Programs for control and prevention of FMD generally apply strategies of diagnostic screening, vaccination, movement restriction, quarantine, and the killing of infected and in-contact animals.

There are a number of disparate factors, however, that can influence the extent to which a country is able to control or eradicate FMD. These factors include technical problems with vaccine efficacy and with diagnostic test accuracy, the absence of political will or economic capacity to provide critical animal health infrastructures necessary to control disease, the geographic proximity of FMD in neighboring countries, and religious and cultural practices and philosophies that prevent application of certain animal control measures. In addition, the nature and biology of the virus imposes other constraints to control. FMD viruses are immunologically diverse, with seven distinct serotypes (A, O, C, SAT1, SAT2, SAT3, Asia1), which can make diagnosis and vaccination problematic. The high mutation rate and massive replication of the virus in infected animals contribute to the extensive genetic diversity of the virus throughout the world. Such diversity can be an impediment to control and eradication of the disease because vaccine strains can become less effective, or ineffective, against newly emerging field strains and because new strains may evade detection if mutations occur at sites of the genome being targeted by diagnostic assays.

A problem faced by all countries, regardless of their FMD status, is that in order for them to be able to prevent or control FMD, they require an awareness and information about the global situation of FMD. Such information would address such questions as where can we expect to find FMD today, and where will it likely be this time next year; what are the conditions necessary for FMD to move from one country to another; what are the new or emerging strains of virus that might not be protected by current vaccine strains or that might not be detected by current PCR assays; what changes have taken place in the projected risk of our country acquiring FMD; what countries or regions should be targeted for additional resources in efforts to stave off predicted new outbreaks and spread.

Increasingly, there is a recognized need for countries and agencies to have situational awareness for FMD and to be able to anticipate new incursions of FMD, of new FMD viruses, or new or elevated risks of FMD so that appropriate measures can be taken in advance to prevent or mitigate disease and its impact. One of the strategies for early detection of and response to FMD is that of surveillance, both at a national and at a global level, that would aim to seek out specific information about FMD, risks of FMD, and the FMD virus that is needed by the international community and by individual countries in planning and preparing FMD programs. A fundamental element of a surveillance system is how well the system can provide information rapidly in order to allow necessary planning and preparation to begin immediately. Thus, it will be critical that a surveillance system route information and analysis of data in real time in order for timely changes in programs to prevent and control FMD.

Although there has been considerable discussion about the needs and prospects for a global surveillance system for foot-and-mouth disease, little in the way of formal action has taken place to create such a system. In this report we describe a new system, referred to as the FMD BioPortal, which is currently operational and aimed at providing real time information, analysis, and visualization of FMD data. We describe new developments in the FMD BioPortal, including key features, its IT design and functionality, and thoughts on future needs for real time surveillance and use of the BioPortal. We also offer some concepts, definitions, and considerations for global FMD surveillance function and operation, which hopefully will encourage advancement of surveillance research and initiation of political will and dialogue necessary to move forward in formalizing international surveillance efforts.

2   Elements and Concepts of Surveillance

2.1   A Working Definition of Surveillance and Other Competing Terms

The definition applied here for surveillance is "an active, ongoing, formal, and systematic process aimed at early detection of a specific disease or agent in a population or at the early prediction of an elevated risk of a population acquiring an infection or disease, with a pre-specified action that would follow detection of the disease, agent, or elevated risk" [1]. A surveillance system can be considered analogous to a diagnostic assay, in which surveillance is applied to a population for the purpose of detecting the targeted agent or disease, if it is truly present (surveillance sensitivity), and of verifying freedom from disease or infection, if it is truly absent (surveillance specificity). The term 'monitoring' is considered here to be a process undertaken to obtain an ongoing assessment of disease trends, and not intended to seek out and find disease. Quite often, programs referred to as surveillance are actually a type of monitoring system. The term 'survey', as used in the context of disease studies, refers to the process followed to identify possible causal factors of disease or to estimate disease prevalence.

2.2   Surveillance System Design and Architecture

Local or regional FMD surveillance design and architecture, particularly that related to sampling schemes, should be directed by the biology and epidemiology underlying FMD in the area, the population dynamics of species susceptible to FMDV infection, and the cultural and social features of the region or country that could influence risk or transmission of infection. The strain-and-host-specific pathogenesis, for example, influences duration of disease transition states, amount of virus shed, severity of clinical signs, and likelihood of transmission to other animals and should be considered in defining sampling schemes. Herd and flock management and husbandry practices can alter the chances for contact between infectious and susceptible animals and some animal trade networks can, more than others, promote spread of FMD to other regions or countries. Cultural or religious practices and events that involve animals also can affect transmission of the virus by bringing together infected and susceptible animals. Surveillance sampling schemes will need to vary depending on the likely location and timing of infected animals, in which high risk animals may be targeted for aggressive, frequent sampling. The design of local or regional surveillance, therefore, should take into account the biology of prevailing serotypes in the host species, as well as the cultural and husbandry practices that will affect changing geographical and temporal distributions of infected animals. Design of a global system also should consider all local, regional, and country-specific systems in a way that would interconnect and support one another, and that would identify in real time the FMD distributions and risk prevailing in the world.

2.3   Surveillance Strategies

We will not discuss the numerous surveillance strategies envisioned for various surveillance objectives, suffice it to say that depending on a country's needs and resources and on the epidemiology of FMD in the region, surveillance objectives and operation would be expected to differ for each country. One possible approach to a global FMD surveillance system could involve a network of surveillance systems, whereby local or regional systems would be nested within a country or sub-continental system, which in turn would be nested within a global framework. The framework would provide the connectivity among the various layers of surveillance; the type or nature of surveillance for each subsystem would depend on whether FMD is epidemic or endemic, or if the area is considered free without vaccination or free with vaccination. An overarching network connecting subsystems would necessarily have to be able to communicate among the dissimilar aspects of each system, including disparate data and reporting formats, different assays and language, and unique aspects of other international animal health agencies, such as those for the Office International des Epizooties (OIE) and the Food and Agriculture Organization (FAO). Thus, design of a global system will need to address methodologies for standardization and translation in order to maximize communication compatibility with previously existing programs and systems.

For many countries that lack adequate infrastructure to detect FMD, let alone develop a surveillance system, international agencies could use surveillance of correlated surrogates for FMD or FMD risk as a proxy for actual disease surveillance. For countries that do not report FMD, but that are known to have the disease, some predictors of FMD presence can include political will and voice and economic capacity [3]. Other information, including OIE data and expert opinion, can be applied in prediction models to obtain estimates of FMD risk in various regions and countries, particularly those where no diagnostic or reporting systems exist [Rebecca Garabed, unpublished data].

Surveillance strategies also will need flexibility to address new or expanded (or diminished) risks and to maximize efficient and effective allocation of resources.

A system that is designed to accommodate a hierarchy of surveillance intensities will permit ramping up or damping down sampling number or frequency, depending on assessment of risk [2]. Thus, surveillance activities and relative allocation of resources should be guided by ongoing risk projections to modify surveillance activities at specific times and geographic locations.

A critical strategic element in a global surveillance system will be real-time information transfer among the various operational groups, including laboratories, field units, and policy and decision makers. Web-based information systems will need to exist and applied in ways that permit easy electronic access to and retrieval of data, information, maps, models, and analyses. The real-time sharing of information will be key to connecting and communicating with operational units, as well as to the early recognition and understanding of emerging risks or changes in the global FMD picture.

2.4 The FMD BioPortal

The FMD BioPortal was developed as a collaborative effort of the Institute for Animal Health (the FMD World Reference Laboratory) at Pirbright, England, the Artificial Intelligence Laboratory at the University of Arizona, and the FMD Laboratory at the University of California, Davis. Version 1.0 was made operational in January, 2007 (see: http://fmd.ucdavis.edu/bioportal/). The initial goal was to create a web-based system that would make FMD-related data that is banked at the Pirbright laboratory available to the public and to those at the laboratory. A primary objective was to be able to apply to the data basic search and analytic tools, including graphic and tabular presentation of the data and cluster analysis, and to be able to download selected records. The data represent cases or outbreaks of FMD for which samples have been submitted to the laboratory in Pirbright since 1957. Generally, sample submissions to the laboratory have been from some OIE-member countries with ongoing programs to control and eradicate FMD. Data available include the outbreak location and time of onset, information about the host or host species, and the serotype of the virus involved in the outbreak. Data are pushed from the Pirbright laboratory to the FMD Lab at UC Davis, and generally the data are 3 months delayed to provide countries sufficient time to address changes in their FMD programs before the data are made available to the public. Version 2.0, which is planned for May 2007, will include two main additions: 1) access to aligned FMD virus sequence data available publicly from GenBank and to tools for real-time development and comparison of phylogenetic trees of virus isolates and 2) historic and current OIE FMD data and tools for graphics and analysis and for downloading the data. The OIE data cannot currently be accessed electronically.

2.5 FMD News

The FMD News is a real time web search service provided by the FMD Lab to identify and capture FMD-related news items appearing worldwide and to direct the information to those interested in global FMD events, directly via email or via the BioPortal. FMD News items represent both official news releases, such as from OIE or governments, and unofficial opinions, commentary, or reports by individuals or the press (see: http://fmd.ucdavis.edu/index.php?id=1). The FMD News offers an opportunity to obtain a global situational awareness of FMD using 'soft' information, as well as some official information. Incorporation of FMD News items into the BioPortal allows tracking, mapping, and management of the information for regions or countries or of specific topics, such as vaccination, trade embargoes, or specific control campaigns.

2.6   FMD BioPortal Design

The FMD BioPortal system was developed onto the previously existing BioPortal platform (see http://www.bioportal.org), which is a general purpose infectious disease information sharing, analysis, and visualization environment [5]. The system architecture of FMD BioPortal is shown in Figure 1. The major systems components of the FMD BioPortal are 1) a communication backbone for secure, real-time data transfer, 2) a data analysis module that can run several analytical programs to assess spatial-temporal clustering (hotspot analysis), and 3) an interactive visualization tool that allows for integrated analysis and display of epidemiological and genomic sequence data.

The communication infrastructure consists of several messaging adaptors that can be customized to interoperate with various messaging systems. Participating FMD surveillance data providers can link to the FMD BioPortal data repository *via* the PHINMS and an XML/HL7 compatible network. In addition to standard table-based data aggregation functions (e.g., based on time intervals

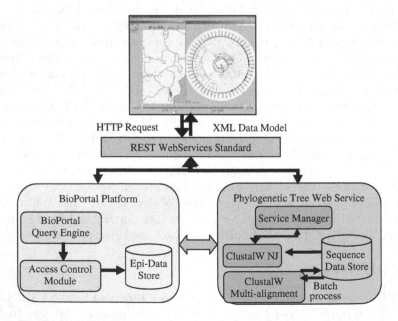

Fig. 1. FMD BioPortal system architecture based on Web Services

and regions), BioPortal has spatio-temporal analysis capabilities to detect spatial and temporal clusters of events or disease, as appropriate for use in disease outbreak investigations. Cluster analysis programs are available to support methods for scan statistic analysis, for risk-adjusted nearest neighbor hierarchical clustering (RNNH), for risk-adjusted support vector clustering (RSVC), and for prospective support vector clustering (PSVC) [4,6,7].

The BioPortal platform has a general-purpose visualization environment, referred to as the Spatial-Temporal Visualizer (STV), which allows users to explore spatial and temporal patterns, based on an integrated tool set consisting of a GIS view, a timeline tool, and a periodic pattern tool [5]. The GIS view displays cases and sightings on a map and the user can select multiple datasets for viewing in different layers, using the checkboxes (e.g., disease cases, natural land features, and land-use elements). Periodic or cyclical temporal patterns of disease also can be viewed. All the functionalities of STV are available through FMD BioPortal to analyze FMD-related epidemiological datasets.

In a new version, FMD BioPortal has applied the STV tool to analysis of virus gene sequence data that have been integrated with epidemiological data, such as time, place, host species, and serotype. The program allows for adjustment of the threshold genetic distance between any two isolates, as chosen by the user, to assess genetic relatedness among FMD virus strains, using a rotary phylogenetic tree display created using the neighbor-joining method. Gene sequence alignment necessary for phylogenetic tree computations is performed using open-source implementations through a web service architecture. This flexible plug-and-play IT architecture enables reuse of tools and leverages use of existing analysis and visualization toolsets.

2.7   Future Display and Analysis Tools

Currently the data captured and displayed by the FMD BioPortal do not represent true surveillance data in the sense that, for the most part, they were not obtained through a formal process to seek out and find the disease. Rather, the data come about as the result of investigations of known or highly suspicious cases of FMD and represent reports of cases submitted by countries with programs to control and monitor the disease. Even though these data may not represent ideal surveillance data, they do lend themselves to scrutiny and analysis that may render some insight into risk or projected risk.

Several tools are being developed or considered for inclusion in the BioPortal that will enhance the ability to detect unusual or unexpected events and to project macro changes in global risk [8,9,10,11,12]. Some of these tools include:

a. Various anomaly detection methods to identify outlier events or cases of FMD, using rule-based anomaly detection, which is under control of the user, and using model-based anomaly detection methods. The latter include various time series and temporal and spatio-temporal models to detect anomalous events or an unpredicted excess (or absence) of disease for a defined area, host, or time period.

b. Prediction models for identification of anomalous or unpredicted genetic variants of the virus that would not be predicted by the evolutionary history for a region and that that might represent new incursions or an 'escape' of virus from another region.
c. Molecular epidemiological models to predict severity, duration, and likelihood of transmission of disease based on molecular changes taking place in the virus over the course of an epidemic.
d. Global prediction models to project changes in risk of FMD in defined geographic grids based on changes in regional economics, trade, and political stability.
e. Models that estimate efficacy of vaccination, using genomic and immunologic information for field strains and for competing vaccine strains.
f. Models that project spread of FMD in a defined region under various control strategies and that can be used in developing disease control programs.
g. Models for surveillance sampling that identify optimal combination of sampling size, frequency, and targeting to maximize the probability of detecting FMD, given the available resources.

2.8   Prospects for International Partnerships for Global FMD BioPortal

Prospects would appear favorable for international collaboration and networking to address global sharing and surveillance for FMD. Creation of a global FMD surveillance network of countries and agencies will require long-term vision, political will, and strong leadership on the part of the U.S. and other countries to provide the best possible program for 1) early diagnosis of FMD and FMD risk, 2) identification of molecular changes in the virus and epidemiological changes in the disease, 3) projection of changes in global risk of FMD, 4) sharing of information, and 5) advancement of the science. Success will be contingent on removing barriers that currently exist for sharing the information needed to achieve common goals for controlling and eradicating FMD globally.

In recognizing that a primary objective in global surveillance will be the real time dissemination and analysis of data for the purpose of making timely decisions, careful attention will need to be given to the operational structure developed for global surveillance through the BioPortal or other web systems. Ideally, the operational organizations should be

a. dedicated primarily to providing rapid and efficient service in disease surveillance to users.
b. efficient, with minimal administrative overhead.
c. able to provide information and analyses in real-time *via* the web.
d. willing and able to foster new develop and to embrace new thinking, ideas, and initiatives.
e. open and unrestricted in that no one group can control the information, and thus can control the science.
f. able to support sustained research and development necessary to improve surveillance and efficiency of operations.

Numerous issues have been raised in discussions about developing a global FMD information system that would share and analyze data through the Bio-Portal, or any other web-based system. Many of these unresolved issues represent roadblocks to progress in moving forward to achieve global surveillance. Some of questions raised include:

a. Who owns the system?
b. Who owns the data?
c. Who is responsible for data integrity?
d. Who controls the data?
e. Who is responsible for maintenance and further development?
f. What agreements are necessary for data sharing?
g. Who should pay?
h. Who should administer and operate the system?

These informal discussions have revealed general agreement that a system such as the BioPortal would greatly enhance efforts globally to reduce risks of FMD and that the current operational FMD BioPortal illustrates techno-logical barriers to development of a real time information sharing system no longer exist. However, these same discussions reveal considerable inertia yet to be overcome internationally in agreeing on how to proceed. Reasons for some of the inaction relate to wariness as to how such a system would serve special geo-political groups. For example, an EU-centric system might encourage a sys-tem that specifically addresses the critical threats the European countries face by their close proximity to FMD; whereas, such as system might not necessar-ily accommodate needs of countries in Southeast Asia. International agencies, such as the OIE or the FAO, may view an FMD surveillance system as falling under their own individual purview. On the other hand, the World Reference Laboratory for FMD at Pirbright has indicated an interest in directing global animal health surveillance in releasing plans develop the ReLAIS web-based sys-tem. Hopefully, U.S. agencies that would want to have global FMD awareness would become more engaged in these discussions.

Now that the FMD BioPortal has moved beyond proof of principle and is operational on some scale, it is time to move forward with long-term planning and development for this system on a multilateral and international level. As a next step in creation of a global network and system for FMD, including the FMD BioPortal, it will be necessary to develop a roadmap for coordinated actions and planning. The following are suggested topics for consideration in developing such a roadmap, perhaps by a consortium of interested international partners and players:

a. Obtain views on global surveillance from the global FMD community.
b. Identify technical issues remaining for data sharing.
c. Identify political barriers for information sharing and possible solutions.
d. Obtain funding for research and development.
e. Encourage and engage stakeholders in development of a global system.
f. Design operational structure for data sharing.
g. Promote participation in the system.

In summary, there are many aspects and issues of global FMD surveillance that were not addressed here and that deserve the benefit of broad debate and discussion. Careful consideration will need to be given at various international levels, which hopefully will include the U.S., for further development of the FMD BioPortal, its tools, and linkages with other Web-based systems that will be necessary to address the needs for research and for prevention, control, and eradication of the disease globally.

The development of the FMD BioPortal represents a critical first step toward realizing a goal of global infectious disease surveillance and in recognizing that global surveillance will not be possible without a system for international real time information sharing about FMD.

Acknowledgments. The authors acknowledge the support for development of the FMD BioPortal provided by a grant from the Armed Forces Medical Intelligence Center and a grant (#IIS-0428241) from the National Science Foundation.

References

1. Thurmond, M.C. Conceptual foundations for infectious disease surveillance. J Vet Diagn Invest 2003;15:501-514
2. Bates T, Thurmond M, Hietala S. Surveillance for detection of foot-and-mouth disease. J Amer Vet Med Assoc 2003;223:609-614
3. Garabed R, Perez A, Johnson W, Thurmond M. Predictive modeling as a tool for global foot-and-mouth disease surveillance. International Symposium of Veterinary Epidemiology and Economics. Cairns, Australia, August 2006.
4. Chang W, Zeng D, Chen H. (2005). "Prospective Spatio-Temporal Data Analysis for Security Informatics," Proceedings of the 8th IEEE International Conference on Intelligent Transportation Systems, Vienna, Austria, September.
5. Hu P J-H, Zeng D, Chen H, Larson C, Chang W, et al (2005). "Evaluating an Infectious Disease Information Sharing and Analysis System," Proceedings of the IEEE International Conference on Intelligence and Security Informatics (ISI-2005), Springer Lecture Notes in Computer Science, Vol. 3495.
6. Zeng D, Chang W, Chen, H. (2004). "A Comparative Study of Spatio-Temporal Hotspot Analysis Techniques in Security Informatics," Proceedings of the 7th IEEE International Conference on Intelligent Transportation Systems, pp. 106–111, Washington, DC.
7. Zeng D, Chen H, Tseng L, Larson C, Eidson M, et al. (2004). "West Nile Virus and Botulism Portal: A Case Study in Infectious Disease Informatics," Intelligence and Security Informatics, Proceedings of ISI-2004, Springer Lecture Notes in Computer Science, Vol. 3073, pp. 28–41.
8. Perez AM, Thurmond MC, Carpenter TE, Grant PW. Use of the scan statistic on disaggregated province-based data: Foot-and Mouth Disease in Iran. Preventive Veterinary Medicine 71:197-207, 2005.
9. Perez AM, Thurmond MC, Carpenter TE. Spatial distribution of foot-and-mouth disease in Pakistan modelled using cattle density estimates. Preventive Veterinary Medicine 76(3-4):280-289, 2006.

10. Thurmond MC, Perez AM. Modeled detection time for foot-and-mouth disease (FMD) virus in bulk tank milk for FMD surveillance. American Journal of Veterinary Research 67(12):2017-2024, 2006
11. Gallego M, Perez AM, Thurmond MC. Temporal and spatial distributions of foot-and-mouth disease under three different strategies of control and eradication in Colombia (1982-2003). Veterinary Research Communications, in press. DOI: 10.1007/s11259-007-0125-1
12. Perez, A., Garabed, R., Kelley, G., Chhetri, B., Valarcher, J.F., Knowles, N., Konig, G., Thurmond, M. UC Davis FMD models for spatial distribution, sampling, and virus evolution. Session of the Research Group of European Commission for Foot and Mouth Disease Control (EUFMD). October 2006. Paphos, Cyprus.

Utilization of Predictive Mathematical Epidemiological Modeling in Crisis Preparedness Exercises

Colleen R. Burgess

MathEcology, LLC, 3120 West Carefree Highway Suite 1-642, Phoenix,
Arizona 85086-9101
Colleen.Burgess@mathecology.com

Abstract. By providing useful measures of the outcome of an influenza pandemic, the utilization of predictive mathematical models can be an extremely valuable tool within crisis preparedness exercises. We discuss our experiences with developing and implementing such a simulation model for use within a regional crisis preparedness exercise, and make recommendations for maximizing the utility of predictive mathematical epidemiological models in general for future exercises.

Keywords: Influenza, predictive model, crisis preparedness exercises.

1 Introduction

Preparedness is essential in order for communities to estimate the potential impact of a disease outbreak, predict the resources needed for effective response and facilitate rapid interventions to minimize morbidity and mortality. It can be difficult, however, to prepare for events with which the given community has had little or no historical experience. This is where exercising potential events can be extremely useful.

1.1 Coyote Crisis Campaign

The Coyote Crisis Campaign (CCC) is a regional crisis response operation designed to exercise local government, military, academic and industry organizations and entities in emergency crisis response. The CCC takes place in Scottsdale, Arizona, and is a year-long planning and learning process resulting in an exercise intended to improve community safety, disaster response training and effective crisis coordination through the development of a planned response and partnerships capable of successfully collaborating in a crisis [2]. The exercise focus for 2007 is a theoretical outbreak of pandemic influenza within the region.

1.2 Predictive Modeling and CCC

Our role in the CCC is to develop and apply a predictive mathematical epidemiological model to assist stakeholders and participants in assessing the impact of pandemic influenza on corporate capabilities and on the regional healthcare system in terms of influenza cases, deaths and vaccinations, as well as personnel and

D. Zeng et al. (Eds.): BioSurveillance 2007, LNCS 4506, pp. 180–189, 2007.

equipment readiness ratings. Special focus has been placed on the interpretation of model parameters and outputs in terms directly applicable to the stakeholders and the region as a whole.

2 Methods

MathEcology was initially invited to participate in the CCC due to an existing application we developed for the United States Department of Defense (DoD). This software application was built upon concepts developed by the United States Centers for Disease Control and Prevention (CDC) for the spread of seasonal flu through the general population. We re-engineered the CDC model for application to a single age-group with multiple functional subpopulations, and variable disease parameters based upon the roles of these subgroups. The general concept of the DoD model is that influenza cases occur during the first wave of a pandemic according to a stepwise approximate normal distribution based upon the anticipated gross attack rate and pandemic duration (see Figure 1); influenza hospitalizations and deaths track along with the time-series of cases according to user-specified parameters for each subpopulation.

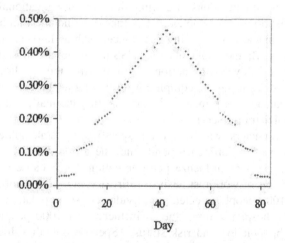

Fig. 1. Simulated proportion of influenza cases among the general population according to a stepwise approximate normal distribution with gross attack rate of 20% and pandemic duration of 12 weeks

We re-structured the model developed for DoD to be applicable for the CCC. The CCC influenza simulation model calculates both time-series and total numbers for influenza cases, deaths, hospitalizations and vaccinations, along with additional outcome measures appropriate to the organizations participating in the CCC. Participation in the numerical simulation activities during the CCC is strictly voluntary, and stakeholders who choose to do so have been provided a user-friendly worksheet to fill in with information regarding their populations of interest, which is then utilized in determining parameter values and ranges for use in the model (see Figure 2 below).

2.1 Parameter Definitions

Information gathered from the stakeholder-completed worksheets is categorized by population parameters, disease parameters and medical resources and equipment.

Each stakeholder can specify an unlimited number of populations to be modeled, and the worksheet fields allow subdivision of populations into groups based on function (for example, soldiers versus airmen), exposure risk, demographics (for example, single employees versus employees with small children) and so on. For each subpopulation additional attributes can be defined, including the percent of the subgroup that are considered critical personnel to the main mission of the organization, the percent that are considered to be of high risk status – having a pre-existing medical condition resulting in a higher risk of contracting an influenza-related illness with a serious health outcome – and the distribution of individuals who are Company Support (CS) and Company Service Support (CSS). The basic mission of CS personnel is to enhance the functionality of the organization by providing critical company support functions in conjunction with organizational employees to secure business operations. Examples of CS functions include training; information systems and communications; enabling assistance, such as nursing staff, interns and assistants; security staff; and public affairs. CSS personnel are those who sustain the ability of company employees to function by transporting the supplies, fuel, food and water; servicing and repairing the equipment; providing health care (in organizations where that is not already the primary mission); sorting the mail; and providing other personnel and administrative services.

The disease parameters which can be specified by stakeholders include the anticipated duration of the influenza pandemic; the gross attack rate, or anticipated percent of clinical cases of influenza per population which cause some measurable company impact (days of work lost, visit to infirmary, etc.); the number of deaths due to influenza per 1000 people infected, subdivided by subpopulation and risk status; and the number of hospitalizations due to influenza per 1000 people infected, also subdivided by subpopulation and risk status. Specific default values for influenza pandemic duration, gross attack rate, death rate and hospitalization rate were determined ahead of time for the CCC by committee, and these values along with recommended values appropriate to seasonal influenza from the CDC are provided in the worksheet for stakeholders that wish to utilize them.

Medical resources such as hospital beds, intensive care unit beds, ventilators and other equipment items can be defined by stakeholders for whom this is appropriate. Participants are also able to specify hypothetical influenza vaccination coverage subdivided according to subpopulation and risk status, in the event that an appropriate vaccine is available at the time of the outbreak. Figure 2 shows a screenshot of the parameters worksheet distributed to CCC participants.

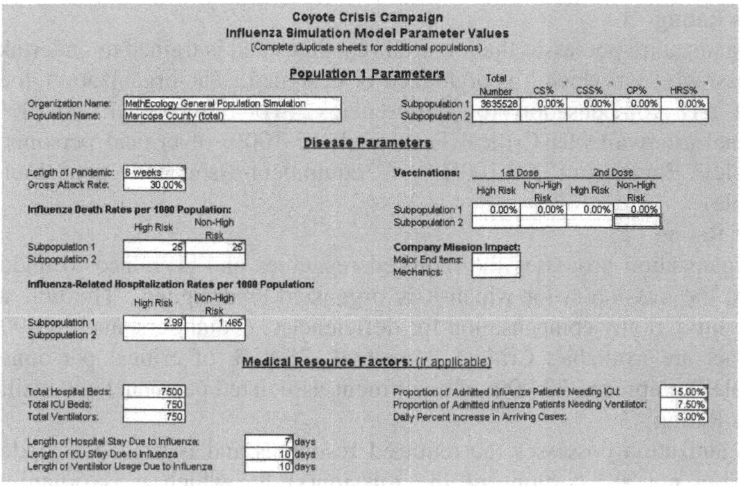

Fig. 2. Sample page from the parameters worksheet distributed to CCC participants, filled in with CCC default values and population parameters for the general population of Maricopa County, Arizona

2.2 Outcome Measures

Aside from the general benefit of estimating influenza morbidity and mortality, the main intent of utilizing predictive models in crisis exercises such as the CCC is to allow organizations to prepare for a crisis event before it arrives and explore organizational vulnerabilities. Thus the outputs provided by the model must offer not just vague numbers but rather focused measures of the impact of pandemic influenza on organizational functioning.

One of the outcome measures provided by the model developed for the CCC is a Readiness Rating – an assessment of the impact of the disease outbreak on organizational equipment (for those stakeholders who provide input values for these fields) or personnel. The Personnel Readiness Rating is based on the relative proportions of personnel and critical personnel (as specified by the user) available to pursue the organizational mission for any given week within the pandemic. The two proportions are compared to scales based on levels of functionality for the given organization, and the most limiting proportion determines the overall Personnel Readiness Rating.

The Equipment Readiness Rating is based on the proportion of equipment-associated personnel (such as mechanics, helicopter-evacuation pilots, or ambulance drivers) available to pursue the organizational mission for any given week within the pandemic. As with the Personnel Readiness Rating, this proportion is compared to a scale based on levels of equipment-associated functionality for the given organization.

The model produces a scaled time-series for the readiness rating on a scale from 1 (optimal) to 4 (inability to function) for each week of the simulated pandemic, with the scales defined as follows:

Readiness Rating - 1

The organization possesses the required resources and is trained to undertake the full mission(s) for which it is organized or designed. The organization does not require any compensation for deficiencies. (Total Personnel: 90-100% of personnel are available; Critical Personnel: 85-100% of critical personnel are available; Equipment: 90-100% of equipment-associated personnel are available)

Readiness Rating - 2

The organization possesses the required resources and is trained to undertake most of the mission(s) for which it is organized or designed. The unit would require little, if any compensation for deficiencies. (Total Personnel: 80-90% of personnel are available; Critical Personnel: 75-85% of critical personnel are available; Equipment: 70-90% of equipment-associated personnel are available)

Readiness Rating - 3

The organization possesses the required resources and is trained to undertake many, but not all, portions of the mission(s) for which it is organized or designed. The unit would require significant compensation for deficiencies. (Total Personnel: 70-80% of personnel are available; Critical Personnel: 65-75% of critical personnel are available; Equipment: 60-70% of equipment-associated personnel are available)

Readiness Rating - 4

The organization requires additional resources or training to undertake its mission(s), but it may be directed to undertake portions of its mission(s) with resources on hand. (Total Personnel: Less than 70% of personnel are available; Critical Personnel: Less than 65% of critical personnel are available; Equipment: Less than 60% of equipment-associated personnel are available)

By evaluating the Personnel Readiness Rating and Equipment Readiness Rating time-series, the stakeholders can determine at what point in the pandemic organizational functioning will be most heavily impacted, and in which areas of functionality. If the equipment readiness rating is predicted to be the limiting factor, then the organization can plan to offer vaccination to medical technicians, mechanics and other equipment-associated personnel in an attempt to circumvent this outcome. If the personnel readiness rating is predicted to be the limiting factor, the stakeholder can then drill down into the time-series output sets for influenza cases and deaths to determine the group hardest hit, and likewise plan to offer prophylaxis to this demographic.

In addition to readiness ratings, the model provides an assessment of the impact of the pandemic on available medical resources such as hospital beds, intensive care unit beds and ventilators at the peak of demand. The hospitalization rate per 1000 cases of influenza as defined by the user in the parameters worksheet, along with relative rates for intensive care unit needs and ventilator usage, is used to calculated the maximum demand on medical resources during the pandemic. This value is compared against the total available hospital beds, ICU beds and ventilators available (for those users who specified these values) to provide the peak proportional utilization for these resources.

3   Results

3.1   Stakeholder Participation

Participation in the numerical simulation activities during the CCC is strictly voluntary. The original parameters worksheet was distributed electronically to 71 individuals from at least 21 organizations at various levels of involvement in the CCC overall, and completed worksheets were returned from five organizations. Of those stakeholders who chose to participate none has elected to model more than three separate populations, and each of these populations is representative of organizational employees as opposed to the community in general.

3.2   Model Results

While specific default values pre-determined for the CCC or recommended by CDC have been provided for disease parameters, most of the respondents chose to utilize their own parameters in at least one of these fields (see Table 1). The resulting model output is thus not directly comparable between stakeholders. While there are specific circumstances under which such disparities are appropriate – for example, emergency room physicians may have a higher gross attack rate than soldiers quarantined to base – in general the value of model output for such tabletop exercises is greatly reduced by the lack of uniformity. In essence, each stakeholder is simulating a different pandemic.

Table 1. Overall specified model parameters for Maricopa County, Arizona and five stake-holders in Scottsdale, Arizona participating in simulation activities for the CCC

Population	Population Size	Pandemic Duration	Gross Attack Rate	Deaths per 1000
Maricopa County, AZ	2,224,943	6 weeks	20%	25.000
Stakeholder 1	849	6 weeks	20%	0.037
Stakeholder 2	2,700	12 weeks	40%	0.037
Stakeholder 3	4,600	12 weeks	40%	25.000
Stakeholder 4	1,234	6 weeks	30%	0.037
Stakeholder 5	6,200	6 weeks	20%	25.000

Table 2. Population distribution for the general population, working adults, school children, seniors and young children and infants living in Maricopa County, Arizona (values based on 2005 census estimates, see [9])

Population	Percentage	Total Number
General Population	100.0%	3,635,528
Working Adults	61.2%	2,224,943
School Children	19.4%	705,292
Seniors	11.1%	403,544
Young Children and Infants (not modeled)	8.3%	301,749

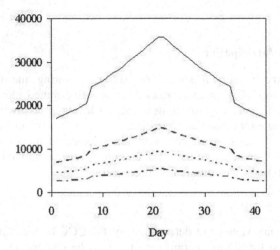

Fig. 3. Predicted time-series of influenza cases for the general population (solid), working adults (dashed), school children (dotted), and seniors (dot-dashed) living in Maricopa County, Arizona under the CCC default assumptions of a six-week pandemic with gross attack rate of 30% for the general population, 20% for working adults, and 40% for school children and seniors

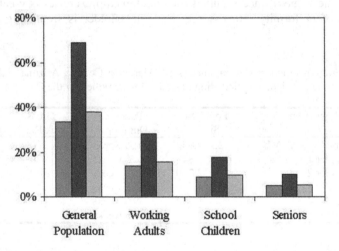

Fig. 4. Predicted utilization of available medical resources for hospital beds (medium grey), intensive care unit beds (dark grey), and ventilators (light grey), based on subpopulation demographics for Maricopa County, Arizona

Of the five stakeholders who chose to participate in the simulation, three specified anticipated vaccination coverages for their populations, while the remaining two stakeholders omitted these values.

In addition to simulations performed according to individual stakeholder preferences, we have also performed a general simulation for the population of Maricopa County, Arizona as a whole and subdivided according to census demographics, shown in Table 2. The predicted distribution of influenza cases during the pandemic, and the utilization of available medical resources for Maricopa County, are displayed graphically in Figures 3 and 4, respectively.

The results of the simulation performed for the general population and subpopulations of Maricopa County will be provided to all stakeholders in the CCC, independent of their participation in individual simulation activities, to offer a general description of the anticipated impact of pandemic influenza on the region as a whole.

4 Discussion

4.1 Vaccination

An additional benefit of utilizing predictive modeling to evaluate the distribution of influenza cases and deaths is to assess priorities in terms of which subpopulations and demographics receive prophylactic vaccination. This has become particularly important in the absence of federal guidance on which groups will receive vaccine if and when it becomes available during a pandemic. Since the availability of sufficient doses of effective prophylactic drugs is questionable at best, it is essential for organizations to have an idea of their vaccine requirements ahead of time. The influenza model developed for the CCC calculates the number of doses of vaccine which must be available in order to implement the specified coverage for first and second doses by subpopulation. At this point in time the model does not calculate the numerical impact of vaccination on the spread of influenza through the organization, nor on the reduction of cases, hospitalizations or deaths, though this valuable enhancement to the model is planned for inclusion in a future application. Nearly half of the respondents opted not to provide vaccination coverage parameters, citing either the belief that in the first waves of a pandemic it is unlikely that an effective vaccine will be developed, much less available for distribution to the general population, or that individuals will obtain prophylaxis through avenues other than their employers.

4.2 Data Sources, Data Quality and Technology Adoption

The CCC is a purely voluntary exercise, and the majority of participants are still required to perform their regular duties in addition to those associated with CCC. As a result, it has been difficult to obtain completed parameter forms by the predetermined deadline. The submitted parameters worksheets were generally submitted from either human resources personnel or individuals representing crisis and emergency management and planning departments. While much of the required information for population parameters is already collected regularly by the human resources departments of the various organizations, particularly information regarding the proportions of company support and company service support, individual organizations have had some difficulty in acquiring appropriate information regarding pandemic duration, gross attack rate, death rate and hospitalization rate, as well as vaccination coverage. Default values for these fields were decided upon for the CCC in general, though the final decision was left to each stakeholder, and as illustrated above each stakeholder has chosen distinct combinations of parameter values, negating the possibility of cross-stakeholder comparison of model results. Some stakeholders left a number of parameter fields blank, and had to be contacted post-submission to verify parameter values before simulations could take place.

There appears to be a general distrust of models and simulations by many organizations – or at the very least, a difficulty in understanding what benefit such activities could possibly offer for the functioning of an organization preparing for an eventual pandemic. We have made an effort to explain the concept and benefits through presentation sessions and handouts, and have made ourselves readily available to answer specific questions from the individuals responsible for filling out parameters worksheets. Our hope is that by providing simulation output for Maricopa County as a whole to all organizations participating in the CCC in a friendly, easy-to-use and helpful format, the value of the output will be immediately evident and these organizations may be more open to the utilization of simulation modeling in crisis preparedness exercises in the future.

4.3 Modeling the Community

While we performed simulations for the impact of pandemic influenza on the general population of Maricopa County based on U.S. Census estimates and without the assistance of local or regional governments, a better way to make the most of the influenza simulation for the purposes of the CCC and crisis exercises in general may be to simulate the outbreak for subpopulations defined by coverage regions for county healthcare facilities.

For such an implementation, the general population would be the county residents, with the various coverage regions for each of the regional healthcare facilities parameterized on separate worksheets similar to those originally distributed to CCC stakeholders. Medical Resources parameters would then be defined for the dominant healthcare facility within each coverage region, as well as vaccination scenarios for the population. General values for gross attack rate, pandemic duration, and so on would be set to CCC defaults to provide uniform and directly comparable results for all regional healthcare facilities.

The output would then be the predicted number of cases, hospitalizations, and deaths within each coverage region, the expected impact on the medical resources of the healthcare facility for that region, and the number of vaccine doses necessary (and time required to give the shots). This information would then be made available to all the stakeholders, whether they submitted individual parameter worksheets or not. Each healthcare facility would then have an idea of the number of patients to expect (in addition to the impact on personnel that would be provided by their individual worksheets). Major employers would have a general idea of family-related worker absenteeism based on the overall percentage of cases among adults and children in the general population. And by evaluating the degree to which demand would exceed available resources, local military resources would have an expectation of if, when and how they would be called upon to assist during the simulated pandemic.

As a result of this kind of implementation the value of mathematical epidemiological modeling would increase, to individual stakeholders and to crisis preparedness exercises in general. Participants in such exercises would be able to come away from the event with more than just practice for possible disease outbreaks, but also information on specific ways in which they can prepare as individuals and as partnerships representing the community as a whole.

Acknowledgments. I am grateful to A. Burgess for critically reading the manuscript, to the anonymous reviewers for their critical review of and recommendations for the improvement of this document, and to the organizing committee of the 2007 Coyote Crisis Campaign for inviting MathEcology to participate in this year's event. The model described in this manuscript is based upon research originally funded by the United States Department of Defense.

References

1. Abbott, A.: What's in the Medicine Cabinet? Nature 435 (2005) 407–409
2. Coyote Crisis Campaign: Purpose and Objectives. Available:
   http://www.coyotecampaign.org/purpobj/ [Accessed 10 January 2007] 2006
3. Ehrenstein, B.P., Hanses, F., Salzberger, B.: Influenza Pandemic and Professional Duty: Family or Patients First? A Survey of Hospital Employees. BMC Public Health 6 (2006) 311
4. Ferguson, N.M., Cummings, D.A.T., Fraser, C., Cajka, J.C., Cooley, P.C., Burke, D.S.: Strategies for Mitigating an Influenza Pandemic. Nature 442 (2006) 448–452
5. Ferguson, N.M., Mallett, S., Jackson, H., Roberts, N., Ward, P.: A Population-Dynamic Model for Evaluating the Potential Spread of Drug-Resistant Influenza Virus Infections During Community-Based Use of Antivirals. Journal of Antimicrobial Chemotherapy 51 (2003) 977–990
6. Longini, I.M., Nizam, A., Xu, S., Ungchusak, K., Hanshaoworakul, W., Cummings, D.A.T., Halloran, M.E.: Containing Pandemic Influenza at the Source. Science 309 (2005) 1083–1087
7. Ridley, R.G.: Research on Infectious Diseases Requires Better Coordination. Nature Medicine Supplement 10(12) (2004) S137–S140
8. Rotz, L.D., Hughes, J.M.: Advances in Detecting and Responding to Threats from Bioterrorism and Emerging Infectious Disease. Nature Medicine Supplement 10(12) S130–S136
9. U.S. Census Bureau: Maricopa County Quick Facts from the U.S. Census Bureau. Available: http://quickfacts.census.gov/qfd/states/04/04013.html [Accessed 11 January 2007] 2006

Ambulatory e-Prescribing: Evaluating a Novel Surveillance Data Source

David L. Buckeridge, Aman Verma, and Robyn Tamblyn

McGill Clinical and Health Informatics, 1140 Pine Ave West, Montreal, Quebec,
Canada, H3A 1A3
david.buckeridge@mcgill.ca, aman.verma@mcgill.ca, robyn.tamblyn@mcgill.ca

Abstract. Researchers have studied many potential sources of data
for biosurveillance but have tended to focus on ambulatory visits and
over-the-counter pharmaceutical sales. Data from electronic prescribing
(e-prescribing) systems in an ambulatory setting have not been evaluated critically, but they may provide valuable data for surveillance. In
this paper we evaluate the utility of e-prescribing data for surveillance
of respiratory infections. Demographic data were analyzed to determine
the differences between patients in an e-prescribing system and the general population. Correlation analysis was performed on the time-series
for common respiratory tract antibiotics and the time-series for respiratory tract infection incidence. Demographic data showed a strong bias
towards older people in the e-prescribing system when compared to the
general population. The analysis also showed that a subset of antibiotics
are highly correlated with respiratory tract indications (0.84, p<0.0001,
95% CI 0.73-0.90). The over-representation of higher age groups in the
electronic prescribing system suggest that data from such systems may
be suitable for observing trends in chronic conditions or infectious conditions more common in the elderly. The results also suggest that a set
of antibiotics can be identified that reflect the incidence of respiratory
tract infections.

1 Introduction

Prescription drugs have been studied previously for their potential value in biosurveillance. In particular, efforts have been made to use prescriptions to detect
reportable diseases. The diseases used in previous evaluation studies were unique
in that the drugs typically prescribed for treatment were used only for the disease in question. Tuberculosis [1,2] and pertussis [3] have been studied in this
way.

When a drug is used exclusively for one medical condition then the prescription rate of that drug is clearly valuable for surveillance of that condition. However, most drugs, especially antibiotics, are used for multiple diseases and this
complicates their use in surveillance.

Some new e-prescribing systems require documentation of a therapeutic indication for each prescribed drug. These indication data allow us to discover

D. Zeng et al. (Eds.): BioSurveillance 2007, LNCS 4506, pp. 190–195, 2007.

what drugs are being prescribed for particular conditions. In this paper, we evaluate the potential utility of data from such an e-prescribing system to determine whether antibiotic prescriptions can be used to conduct surveillance for respiratory infection.

2 Methods

2.1 e-Prescribing System

We used data from the MOXXI (Medical Office of the XXIst Century) e-prescribing system for this study. MOXXI was developed as a portable electronic prescribing system for primary-care physicians to write and transmit prescriptions. The performance, acceptability and use of the MOXXI system have been well-established [4].

The MOXXI system requires that a therapeutic indication be entered by the physician at the time of prescribing. The therapeutic indication may be entered as free-text, or it may be picked from a list. The list of therapeutic indications is developed and maintained by a private company (Vigilance Santé). The list is produced by combining the therapeutic indications listed on the drug monographs with common off-label uses of the drug determined through ongoing literature review.

2.2 Study Participants

One year of data was used in this study to allow observation of seasonal variation in prescriptions and therapeutic indications (May 1st, 2005 - April 30th, 2006).

The 90 physicians that participated continuously in MOXXI during the studied time interval were considered the *study physician group*. A physician was considered to be participating in MOXXI if they had signed the participation form before May 1st, 2005 and were using the system regularly during the following year.

In order for a physician to prescribe through the MOXXI system, patient consent is required. The consent form authorizes the provincial health insurance board (Régie de l'assurance maladie du Québec: RAMQ) and the provincial health and social services ministry (Ministère de la santé et des services sociaux: MSSS) to transmit medical claims and prescription data to the MOXXI research team.

The *MOXXI patient cohort* was defined as the set of patients that were consented through at least one of the study group physicians and had at least one prescription made through the MOXXI system by one of the study group physicians during the interval May 1st, 2005 - April 30th, 2006. The physicians progressively consented more patients during the study, and thus the size of the MOXXI patient cohort increased over the study interval. At the start of the interval the MOXXI patient cohort comprised 7,131 patients, while at the end it comprised 18,852 patients.

2.3 General Population

In order to verify the generalizability of the results, we compared the demographic profile of the MOXXI patient cohort to that of the general patient population of the study physician group. The general patient population was defined as any patient who had received a chargeable medical service from one of the selected physicians within the interval of May 1st, 2005 to April 30th, 2006.

The medical service claims database from RAMQ was used to determine the population of patients who had received a medical service through public insurance over this interval from one of the study physician group. The RAMQ beneficiary database provided data on age and sex of this general patient population.

2.4 Drug and Indication Selection

Three therapeutic indications (bronchitis, pharyngitis, and pneumonia) were selected to estimate bacterial respiratory tract infections. In order to determine the drugs most likely to predict the incidence of respiratory tract infection the top three drugs utilized for these indications were selected (azithromycin, moxifloxacin, and clarithromycin). A weekly aggregated time-series was created for both the common respiratory tract antibiotics, as well as their therapeutic indications, normalized by the number of patients in the cohort that month.

3 Results

Patients in the MOXXI cohort tended to be older than the general population, which might suggest a physician bias in who is selected for MOXXI consent (Table 1). In all age categories from 40-79, MOXXI has a higher proportion of patients. The most striking difference, however, is in the 0-9 age range, 2.9% of MOXXI patients were in this age group, as compared to 13.9% in the general population. There was also a slight tendency towards more female patients in the MOXXI cohort when compared to the general population (Table 2).

There were 78,892 prescriptions (each with an associated therapeutic indication) recorded for the MOXXI cohort patients over the studied interval. 1,074 (1.4%) of these prescriptions were for respiratory tract infections. 1,000 (1.3%) of these prescriptions were for azithromycin, moxifloxacin, or clarithromycin. The specificity of the three drugs for respiratory tract infections is high (Table 3) due to the low prevalence of non-respiratory indications for which the three drugs are prescribed. The sensitivity of all three drugs together is high (53.3%) when compared to the sensitivity of any one drug (20.9%, 17.9%, 14.5%).

The time-series for respiratory tract indications peaks during the winter months, following the expected seasonal trend (Figure 1). The weekly-aggregated time-series for the three antibiotics and the three therapeutic indications were strongly correlated with the weekly trend in respiratory infections (0.84, $p<0.0001$, 95% CI 0.73-0.90).

Table 1. The age distribution of the MOXXI cohort and the general population are compared. The table has been normalized by the number of patients and shows only the proportion of patients in a given age range.

Age	MOXXI	General
80+	0.017	0.017
75-79	0.042	0.024
70-74	0.066	0.033
65-69	0.084	0.042
60-64	0.090	0.046
55-59	0.103	0.063
50-54	0.120	0.080
45-49	0.115	0.086
40-44	0.100	0.088
35-39	0.075	0.085
30-34	0.047	0.068
25-29	0.035	0.059
20-24	0.034	0.063
15-19	0.026	0.058
10-14	0.017	0.048
0-9	0.029	0.140

Table 2. The gender distribution of the MOXXI cohort and the general population are compared. The table has been normalized by the number of patients.

Sex	MOXXI	General
Female	61.0%	58.2%
Male	39.0%	41.6%
Missing Data	0.0%	0.3%

Table 3. The sensitivity, specificity, and positive predictive value rates for all, or one of the three common upper-respiratory tract infection drugs in predicting a therapeutic indication of a common upper-respiratory tract infection. (PPV - Positive Predictive Value)

	Sensitivity	Specificity	PPV
Azithromycin	20.9%	99.9%	71.2%
Moxifloxacin	17.9%	99.7%	50.7%
Clarithromycin	14.5%	99.9%	51.1%
All Drugs Combined	53.3%	99.4%	57.3%

4 Discussion

Electronic prescribing systems are used increasingly by physicians. In this study we demonstrated the potential utility of the data captured in these systems for surveillance. In particular, our results suggest that therapeutic indications for respiratory infections follow the expected seasonal pattern, with lower incidence

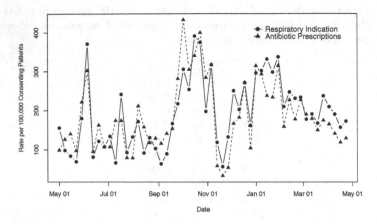

Fig. 1. The time-series of the top three drug prescriptions and the time-series for respiratory infection indication

in summer months and higher incidence in winter months. We also have demonstrated that a set of antibiotics used for respiratory infections, as well as other conditions, follows the same seasonal trend as the indications for respiratory infections.

While our results demonstrate the potential utility of e-prescribing data for surveillance, we also noted important biases in the use of e-prescribing. Most notably, physicians tended to use the e-prescribing software more for elderly patients and females, while the software was used infrequently for young children. These biases could prove important in surveillance. For some surveillance applications the over-representation of elderly may prove beneficial, while in other applications the under-representation of young children may prove problematic, especially for respiratory infections.

The low prevalence of respiratory tract indications recorded in the e-prescribing software suggest that MOXXI isn't used heavily for acute conditions. One possible reason for this is that physicians may find the electronic prescription system and its associated decision support functions more useful in patients with chronic conditions and multiple concurrent prescriptions. The strong bias towards older people in the MOXXI system support this hypothesis.

The interval over which the therapeutic indications and prescriptions were studied was relatively short (1 year), and in future work we will use a longer interval to allow more detailed seasonal analysis.

Finally, while the therapeutic indications given in MOXXI were picked by physicians from a list, there has been no validation of these indications against medical charts. Future work may include studying the validity of theraoeytuc indications through chart audit and the ICD-9 code given in the medical service database.

5   Conclusions

The results of this study suggest that electronic prescribing systems may be useful for the surveillance of respiratory infections. Therapeutic indications for respiratory infections follow the expected seasonal trend and a set of antibiotics predict these indications accurately. However, biases in the population for which physicians use e-prescribing may impact the utility of this data source for some types of surveillance.

References

1. Maggini M., Salmaso S., Alegianti S.S., et al. Epidemiolgoical use of drug prescriptions as markers of disease frequency : an Italian experience. Journal of Clinical Epidemiology. **44** (1991) 1299–307.
2. Yokoe D.S., Subramanyan G.S., Nardell E., et al. Supplementing tuberculosis surveillance with automated data from health maintenance organizations. Emerging Infectious Diseases **5**(1999) 779–87.
3. Chen J.H., Schmit K., Chang H., Herlihy E., Miller J., Smith P. Use of Medicaid prescription data for syndromic surveillance–New York. Morbidity and Mortality Weekly Report 54 (2005):31–4.
4. Tamblyn R., Huang A., Kawasumi Y., et al. The development and evaluation of an integrated electronic prescribing and drug management system for primary care. Journal of the American Medical Informatics Association. **13(2)** (2006) 148–59.
5. Pavlin J.A., Murdock P., Elbert E., et al. Conducting population behavioral health surveillance by using automated diagnostic and pharmacy data systems. Morbidity and Mortality Weekly Report **53** (2004):166–72.

Detecting the Start of the Flu Season

Sylvia Halász[1], Philip Brown[1], Colin R. Goodall[1], Arnold Lent[1], Dennis Cochrane[2], and John R. Allegra[2]

[1] AT&T Labs – Research
[2] Emergency Medical Associates of New Jersey Research Foundation

Abstract. We have combined two methods to detect anomalies in a time series – in this case in emergency department visit data. The n-gram method applies an existing ICD classifier to a set of emergency department (ED) visits for which both the chief complaint (CC) and ICD code are known. A collection of CC substrings (or n-grams), with associated probabilities, are automatically generated from the training data. This information becomes a CC classifier which is then used to find a classification probability for each patient. The output of this classifier can be used to build volume predictions for a syndromic group or can be combined with a selected threshold to provide syndromic determinations on a per-patient basis. Once the daily volume predictions have been calculated using the n-grams, the HWR anomaly detection algorithm is applied, which alerts both for unusual values and for changes in the overall behavior of the time series in question. The earliest alert was generated by the series of volume predicted by flu n-grams as a proportion of total daily visits.

1 Introduction

Syndromic surveillance ED visit data is often based on computer algorithms that assign patient CC to syndromes. ICD9 code data may also be used to develop visit classifiers for syndromic surveillance, but the ICD9 code is generally not available immediately, thus limiting its utility. A number of different methods are currently used to classify patients into syndromic groups based on the patient's CC. In [3] we had reported results using an "n-gram" text processing program adapted from business research technology (AT&T Labs). The method applies the ICD9 classifier to a training set of ED visits for which both the CC and ICD9 code are known. A computerized method is used to automatically generate a collection of CC substrings with associated probabilities, and then generate a CC classifier. Previously, we presented data showing good correlation between daily volumes as measured by the n-gram and ICD9 classifiers.

Detection of the start of the flu season by syndromic surveillance of ED visit data may be complicated, as physicians may be reluctant to diagnose a patient with the flu unless there is news of the flu in their community. Thus a patient presenting with a fever or respiratory illness may be the first harbinger of the flu, rather than a patient given the diagnosis of flu.

D. Zeng et al. (Eds.): BioSurveillance 2007, LNCS 4506, pp. 196–201, 2007.

2   Objective

Our goal was to compare the ability of the following nine classifiers for detecting the outbreak of the flu: flu ICD9, flu n-grams, fever ICD9, fever n-grams, Respiratory (RESP) ICD9, RESP n-grams and the proportion to all daily visits of flu n-grams, fever n-grams and RESP n-grams.

3   Methods

We used a computerized database of consecutive visits seen by ED physicians from 1996 to 2006. We used as our ICD9 classifiers the ICD9 code 487 for flu, and existing ESSENCE filters for fever and RESP. The ICD9 classifiers were applied to a training set of over 1.1 million visits for the period January 1, 2005 to July 31, 2006 to create the n-gram based CC algorithms. We then used the n-gram CC and ICD9 classifiers to estimate the daily volume of patients ending up with and ICD9 code of flu for a test set of 623,787 visits (June 1, 2003- May 31, 2004).

The CC classifiers offer an advantage in that the chief complaints are generally available 5 days earlier than ICD9 codes for most emergency departments. The n-gram method offers a different approach to using manual and natural-language processing techniques to create CC classifiers. It has the advantages that it allows the rapid automated creation and updating of CC classifiers based on ICD9 groupings and may be independent of the spoken language or dialect.

We used the daily totals generated by the different classifiers to generate time series: one using the ICD9 classifier and one series using the n-grams. We also created a third time series for each classifier by dividing the expected daily visits from the n-gram methods by the total daily visits to obtain the proportion of the n-gram totals to the total number of visits. We then used the HWR anomaly detection algorithm to detect outbreaks.

HWR differs from the usual Holt-Winter method in the following:

a.   We are not using forecasted daily values, but residuals based on differences of moving averages.
b.   The seasonality is taken into account in estimating the variances.
c.   Having created the standardized residuals, statistical process control techniques are applied.

To give some statistical characterization of HWR, we have generated a time series using a random Poisson process with a given parameter λ, then applied the HWR algorithm with the rather insensitive setting of the parameters used in the flu study. The algorithm has generated alerts on a number of days. All these alerts are considered false positives, since there were no real outbreaks in the data.

We define the baseline specificity of our alerting algorithm the following way:

Let FP = number of days with alerts; the number of true negatives, TN = total number of days − FP. The baseline specificity will be Spec = TN / (TN + FP).

We have also added simulated outbreaks into the random Poisson process with a maximum of int (α (Q3 − Q1) + 0.5 (Q3 − Q2) + 0.5), where Q1, Q2 and Q3 are the quartiles of the original distribution and int takes the integer part of the quantity. The algorithm could handle non-integers, but we wanted the data to look realistic in a public health framework. The outbreaks were of type 1 in H. Burkom's list of epicurves for $\lambda \leq 100$ and of type 2 for $\lambda \geq 200$ [1] with $\alpha = 1.75$ for $\lambda \leq 10$, $\alpha = 3.5$ for $\lambda \geq 100$. The resulting sensitivities (SENS) and specificities (SPEC) of HWR with the parameter settings used for the flu data are shown in Table 1 together with the baseline specificities.

We are aware of the fact that this pure Poisson data set doesn't present several of the difficulties arising in real time series, but at least gives us some theoretical measures of the algorithm. In [4] we have presented a more-detailed study of the specificity/sensitivity trade-off of the HWR algorithm, using simulated outbreaks with real ED data.

Table 1. Behavior of the HWR algorithm for the parameter settings used in this study

λ	1	5	10	100	200	300
α	1.75	1.75	1.75	3.5	3.5	3.5
Baseline specificity	0.968	0.963	0.970	0.993	0.997	0.997
SPEC with outbreaks	0.964	0.957	0.966	0.993	0.997	0.996
SENS with outbreaks	0.973	0.856	0.813	0.867	0.853	0.840
Maximum of outbreak	4	6	8	49	79	109
Days in outbreak	4	4	4	4	6	6

4 Results

For this study we set the parameters of HWR by constraining it to at most five alerts outside of the flu season defined by the clinicians. We considered all signals out of the flu season to be false positives. Two clinicians by consensus selected the beginning and ending dates for the flu season by analyzing a time series graph of the daily flu visits as determined by the ICD9 flu classifier. For each of the classifiers we recorded the first alert occurrence around the start of the flu season.

Table 2. Alerts for the different time series in a two-week time period

Time series	First alert in flu season	No. of alerts outside the flu season	11/17				11/21 = Start			11/24						12/1
Flu ICD9 totals			0	0	0	0	2	2	4	0	0	2	1	6	7	8	4
			Alerts based on the given time series													
Flu ICD9	11/21	5					X									
Flu n-gram	11/24	0								X						
Flu n-gram proportion	11/22	1							X							
Fever ICD9	11/29	1												X		
Fever n-gram	12/1	1	X													X
Fever n-gram proportion	11/24	1	X							X						
RESP ICD9	12/10	1														
RESP n-gram	12/11	1														
RESP n-gram proportion	12/15	5														
Flu ICD9 with EWMA2	11/23	2							X							
Flu n-gram with EWMA2	11/30	0													X	

The flu n-gram proportion classifier had the earliest signal within the flu season. The fever n-gram and the fever n-gram proportion could be considered leading indicators for this outbreak, since both alerted a few days before the official start. Table 2 shows the 15 day period around the beginning of the flu season. At the bottom are results using EWMA2 from H. Burkom's spreadsheet for comparison.

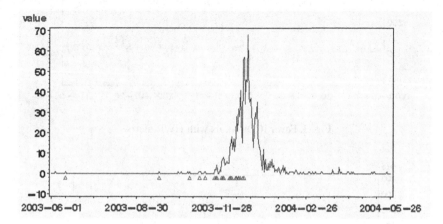

Fig. 1. Flu ICD9 totals with HWR alerts

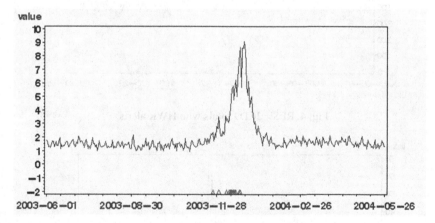

Fig. 2. Flu n-gram totals with HWR alerts

Figures 1 - 4 show times series for the daily visits for the flu, flu n-gram daily expected values, fever and RESP with their own alerts. Figure 5 shows the flu totals based on ICD9 with alerts based on the three n-gram totals: flu n-grams, fever n-grams and RESP n-grams. Figure 6 shows the flu totals based on IC9 with alerts based on the three proportions of n-gram totals to the total number of patients seen that day: flu n-gram proportion, fever n-gram proportion and RESP n-gram proportion.

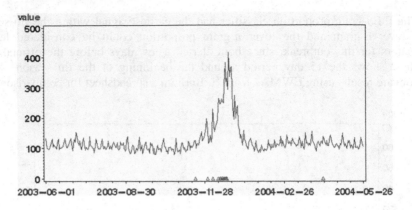

Fig. 3. Fever ICD9 totals with HWR alerts

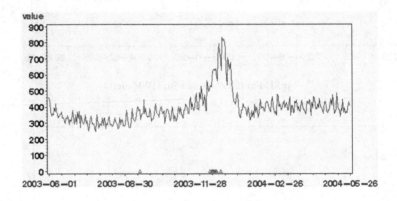

Fig. 4. RESP ICD9 totals with HWR alerts

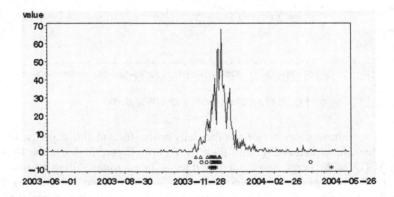

Fig. 5. Flu by ICD9, alerts based on n-grams : triangle=flu, circle=fever, star=RESP

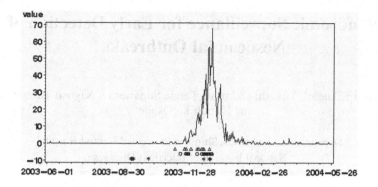

Fig. 6. Flu by ICD9, alerts based on n-gram proportions: triangle=flu, circle=fever, star=RESP

Conclusion. We found that the proportion of flu n-grams to the total number of patients had the earliest signal in the flu season (as defined by the clinicians). The comparison was made between the classifiers by constraining the number of false positives to be less than or equal to 5. This was possible by adjusting the parameters in the HWR algorithm. HWR alerted a few days earlier than the EWMA2 method with comparable false positive rates.

References

[1] H. S. Burkom: Accessible Alerting Algorithms for Biosurveillance. 2005 National Syndromic Surveillance Conference
[2] C. Chatfield & al.: A new look at models for exponential smoothing. The Statistician (2001), 141-159.
[3] S. Halász, P. Brown, C. Goodall, D. G. Cochrane, J. R. Allegra : The n-Gram CC Classifier: A Novel Method of Automatically Creating CC Classifiers Based on ICD9 Groupings. In Advances in Disease Surveillance (2006).
[4] S. Halász, C. Goodall, A. Lent, J. R. Allegra, D.G. Cochrane: An Adaptive Anomaly Detection Algorithm. ISDS Annual Conference (2006)

Syndromic Surveillance for Early Detection of Nosocomial Outbreaks

Kiyoshi Kikuchi[1], Yasushi Ohkusa[2], Tamie Sugawara [2], Kiyosu Taniguchi[2], and Nobuhiko Okabe[2]

[1] Department of Pediatrics, Shimane Prefectural Central Hospital
[2] National Institute of Infectious Diseases

1 Objective

Syndromic Surveillance is typically a system used for early detection of bioterrorism attacks, pandemic flu or other emerging diseases, which monitors symptoms of outpatients or is conducted in the Emergency Department. However, if we monitor symptoms of inpatients, we can apply Syndromic Surveillance to early detection of nosocomial infection. To test this possibility, we constructed and are performing a Syndromic Surveillance System for inpatients who have fever, respiratory symptoms, diarrhea, vomiting or rash. We will then evaluate its statistical properties and its usefulness.

2 Method and Material

With the cooperation of a large hospital which has utilized electronic medical records since August 1999, we use the data they have collected of the number of inpatients who have a certain type of symptom. So as to detect nosocomial outbreaks ward by ward, we have to use the number of patients in the same ward who share a certain symptom over the total number of patients who have the same symptom as a monitoring variable. In order to detect outbreaks, we at first estimate the baseline using the data from August 1st, 1999 to the day before any given day. Then we predict the number of patients in that day and judge whether or not an outbreak has occurred. We use ordinary least square estimation to estimate a baseline which contains dummy variables for the epidemiological week number, the day-of-the-week, national holidays, the day after national holidays and long term trends as explanatory variables. The estimation equation is;

The number of cases of a symptom i in ward j on day t / The total number of cases with the same symptom in all inpatients

$$=\alpha^{ij}+\Sigma_k\beta^{ij}_k \text{ (Week No)}_{kt}+\Sigma_k\gamma^{ij}_v \text{ (Day-of-the-Week)}_{kt}$$

$$+\eta^{ij} \text{ (the Day after a holiday)}_t+\theta^{ij}t+\delta^{ij}t^2+\varepsilon_t$$

Surveillance systems must be evaluated in terms of timeliness, sensitivity and specificity. Usually, a gold standard is defined and we check to see how the

D. Zeng et al. (Eds.): BioSurveillance 2007, LNCS 4506, pp. 202–208, 2007.

surveillance system differs from it. The gold standard for a detection algorithm in this system would be a laboratory confirmed nosocomial outbreak, though this may be a rare event. However this hospital did experience a laboratory confirmed nosocomial outbreak of the Noro virus on January 27th, 2005. We will check the performance of this system by using this confirmed outbreak as a gold standard.

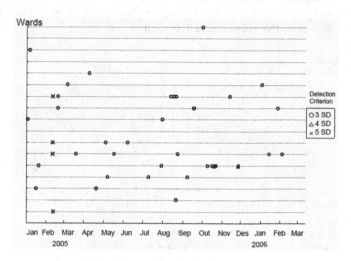

Fig. 1. Fever

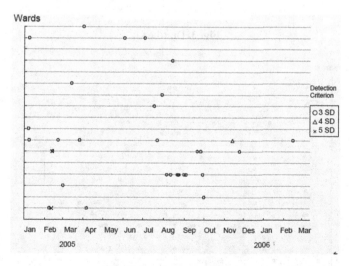

Fig. 2. Respiratory symptoms

Additionally, we also perform an evaluation through computer simulation as in other studies of Syndromic Surveillance. This simulation is performed by adding x cases to the observed data in each day as noise or a nosocomial outbreak. If x is large, say 5, 10, or 15, since it indicates that it is irregular and may be a nosocomial outbreak, the system makes an alert. Therefore, sensitivity is defined as (the number of alerts / the number of simulations). Conversely, if *x* is small, say 1, 2, or 3, since it is a usual event, the system should not detect a nosocomial outbreak. Therefore, specificity is defined as 1-(the number of alerts / the number of simulations).

We assume three criterions for outbreak detection, i.e. three standard deviation of residuals, four standard deviation, and five standard deviation. If the number of patients sharing the same symptom within a ward is higher than the baseline by more than these criterions, we recognize that there is an aberration.

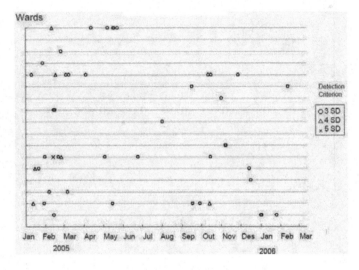

Fig. 3. Diarrhea

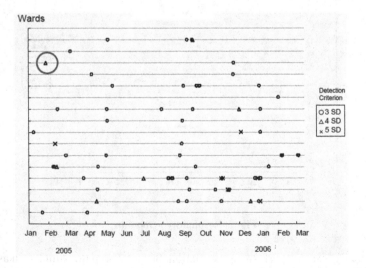

Fig. 4. Vomitting

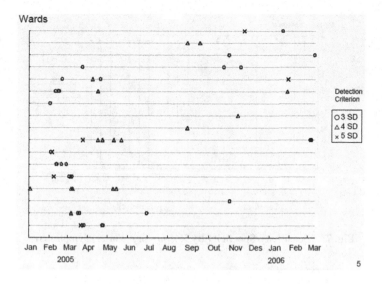

Fig. 5. Rash

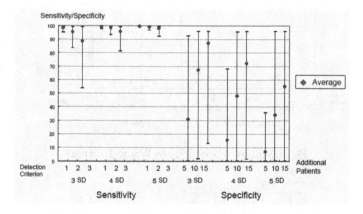

Fig. 6. Sensitivity/Specficity by Wards(Fever)

3 Results

Figures 1-5 show the dates of detected outbreaks of fever, respiratory symptoms, diarrhea, vomiting and rash, by ward. The 17 horizontal lines indicate each ward, while the crosses indicate outbreak detection with the highest criterion, the triangles are for moderate level, and the circles are for lower level. The big red circle in Figure 4 shows the ward where the nosocomial outbreak of the Noro virus was confirmed on January 27th, 2005. This system found an outbreak of vomiting at moderate criterion on this

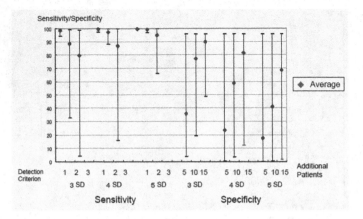

Fig. 7. 別 Sensitivity/Specficity by Respiratory Wards (Symptoms)

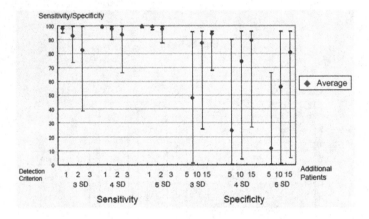

Fig. 8. Sensitivity/Specficity by Wards (Diarrhea)

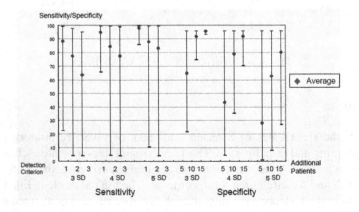

Fig. 9. Sensitivity/Specficity by Wards (Vomitting)

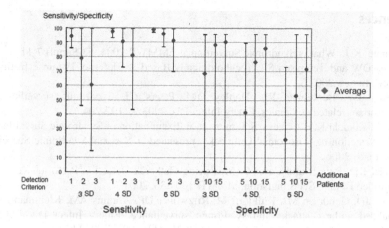

Fig. 10. Sensitivity/Specficity by Wards (Rash)

day. Figures 6-10 show the distribution and averages for sensitivity and specificity among wards.

4 Discussion and Conclusion

Using two different approaches we confirmed that this system can detect nosocomial outbreaks. One approach uses a confirmed nosocomial outbreak and the other is through computer simulation. The system was able to detect the confirmed nosocomial outbreak at the moderate alert level. However, the computer simulation shows a large difference in sensitivity and specificity among hospital wards. Namely, in wards where patients with a certain symptom are rare, it enjoys high sensitivity, but has low specificity. Conversely, in the wards where patients with a certain symptom are common, it suffers from low sensitivity, but has high specificity. Such characteristics of these wards should be removed through adjustment using some explanatory variables, but this remains as further research.

On August 1st, 2006, we started using an automated system; data collection, statistical analysis for detecting clusters, and sending e-mail to members of the infection control team are all completely automatic functions. The infection control nurses then confirm whether there is a true nosocomial infection; by checking electronic medical records, asking other nurses or doctors about patients, or observing the patients themselves.

Currently, we have reformed the system to exclude cases in which the patients had a certain type of symptom when they were admitted and that symptom has not been cured, because this would be an infection from outside the hospital and not a nosocomial one. Within the past two months, we had only one high level alert, but we can confirm that this event was not due to infection.

References

1. Henning. K.J., What is Syndromic Surveillance ?.MMWR 2004; 53(Suppl):7-11
2. Siegist DW and Tennyson SL, Technologically-Based Biodefense, Potomac Institute fro Policy Studies, 2003.
3. Buehler JW, Berkelman RL, Hartley DM, Peters CJ. Syndromic surveillance and bioterrorism-related epidemics. Emerg Infect Dis. 2003:9;1197-204
4. Kiyoshi Kikuchi K,Y Ohkusa, S Tammy et al, Examination of Syndromic Surveillance for Early Detection of Nosocomial Outbreaks, presented at National Syndromic Surveillance Conference 2006.
5. Mandel KD, Reis B and Cassa C. Measuring Outbreak-Detection Performance by using Controlled Feature Set Simulation, MMWR 130-136, 2004.
6. Nordin JD, Goodman MJ, Kulldorff M, Ritzwoller DP, Abrams AM, Kleinman K, et al. Simulated anthrax attacks and syndromic surveillance. Emerg Infect Dis. 2005 Sep. Available from http://www.cdc.gov/ncidod/EID/vol11no09/05-0223.htm
7. Buckeridge DL, Burkom H, Moore A, Pavlin J, Cutchis P, Hogan W. Evaluation of syndromic surveillance systems design of an epidemic simulation model. MMWR 2004; 53(Suppl):137-43.
8. Kulldorff M, Zhang Z, Hartman J, Heffernan R, Huang L, Mostashari F. Benchmark data and power calculations for evaluating disease outbreak detection methods. MMWR 2004;53(Suppl):144-51.
9. Hutwagner L, Thompson W, Seeman GM, Treadwell T. The bioterrorism preparedness and response Early Aberration Reporting System (EARS). J Urban Health. 2003;80: 89-96.
10. Hutwagner L, Browne T, Seeman GM and Fleischauer AT. Comparing Aberration Detection Methods with Simulated Data, Emerging Infectious Diseases, 2005 11(2),314-316.

A Bayesian Biosurveillance Method That Models Unknown Outbreak Diseases

Yanna Shen and Gregory F. Cooper

Department of Biomedical Informatics and the Intelligent Systems Program
M-183 VALE, 200 Meyran Ave
University of Pittsburgh, Pittsburgh PA 15260
{shenyn,gfc}@cbmi.pitt.edu

Abstract. Algorithms for detecting anomalous events can be divided into those that are designed to detect specific diseases and those that are non-specific in what they detect. Specific detection methods determine if patterns in the data are consistent with known outbreak diseases, as for example influenza. These methods are usually Bayesian. Non-specific detection methods attempt broadly to detect deviations from some model of the non-outbreak situation, regardless of which disease might be causing the deviation. Many frequentist outbreak detection methods are non-specific. In this paper, we introduce a Bayesian approach for detecting both specific and non-specific disease outbreaks, and we report a preliminary study of the approach.

Keywords: anomaly detection, biosurveillance, Bayesian methods.

1 Introduction

Detection of anomalous events in data has important applications in domains such as disease outbreak detection [1], fraud detection [2] and intrusion detection [3]. In a typical scenario, a monitoring system examines a sequence of data to determine if any recent activity can be considered a deviation from baseline behavior. These anomalous events can be divided into two types – those that we know about and those that are unexpected. As a result, algorithms within these monitoring systems can be classified into two categories that we will refer to as specific detection algorithms and non-specific detection algorithms. A robust detection system would use a combination of detection algorithms from both of these categories.

Specific detection algorithms look for pre-defined anomalous patterns in the data. For example, in the context of disease-outbreak detection, a specific detection approach might examine health-care data for the onset of a particular disease, such as inhalational anthrax. In contrast, a non-specific detection approach would try to detect any anomalous events that are missed by the specific detectors. By combining these two approaches we might be able to obtain a hybrid approach that detects anticipated diseases well, while having the non-specific approach serve as a "safety net" that is able to detect unanticipated (and possibly never before seen) diseases. We call this combined approach a *safety-net algorithm*. In this paper, we describe a Bayesian safety-net algorithm for detecting disease outbreaks. While the analysis in this paper

D. Zeng et al. (Eds.): BioSurveillance 2007, LNCS 4506, pp. 209–215, 2007.

is basic, we can apply the fundamental ideas to develop much richer Bayesian safety-net models and detection algorithms.

2 Methodology

This section introduces an example model and describes how we use the model for outbreak detection. Due to space limitations, we are only able to present an outline of the complete method.

Let d_0 represent all the diseases that Emergency Department (ED) patients can have in the absence of any disease outbreak in the population. Let d_k denote one specific outbreak disease that we know about. If we assume that there are n types of outbreak diseases, then $1 \leq k \leq n$. Finally, we use d^* to represent any unknown or unexpected outbreak disease.

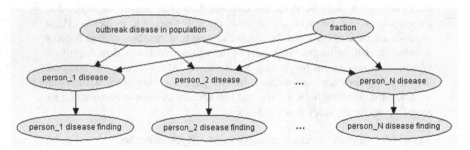

Fig. 1. A Bayesian network showing the population-wide disease model

The disease detection model we use is a population-wide Bayesian network model, as shown in Fig. 1, which represents all the people in the population (not just the ED patients). Let the total number of the population be N. Let i represent the index of a specific person in the population, where $1 \leq i \leq N$. We use pop_dx to represent the values that node *outbreak disease in population* can take. Then pop_dx could be *NOB* (no outbreak), d_k (outbreak of known disease d_k) or d^* (outbreak of unknown disease d^*). The node *fraction* represents the fraction of the total population who have the outbreak disease (either d_k or d^*) and visit the ED. We use psi_dx to denote the values that node *person_i disease* can take, which represents the possible diseases that person i can have given pop_dx. For the people who do not come to the ED, we assign them the disease state called *none*. For a patient who comes to the ED, his or her disease state is a latent (hidden) variable.

When $pop\_dx = d_k$ (or d^*), a specific person i could have disease d_0, d_k, d^* or *none*. The probability of person i having d_k (or d^*) is equal to the value of the *fraction* node by the construction of that node. If $pop\_dx = NOB$, a specific person i either has d_0 or he/she did not come to the ED, and therefore $pop\_dx = none$. The probability that the person has d_0 is estimated from past ED data.

Given the disease state of person i, as represented by psi_dx, we use psi_fd to model the state of a binary symptom of that person. The symptom state of a person is modeled using a Bernoulli distribution. It is possible to model more than one symptom, but for simplicity of presentation, we restrict this paper to an example that contains only one symptom. In particular, we consider the symptom states as being *cough* or *no cough*. For the patients that come to the ED, we define $P(ps_i\_df = cough \mid ps_i\_dx = NOB) = p_0$, $P(ps_i\_df = cough \mid ps_i\_dx = d_k) = p_k$ and $P(ps_i\_df = cough \mid ps_i\_dx = d^*) = p^*$. In the next section we describe how we model p_0, p_k and p^* under uncertainty.

2.1 The Disease Model

As stated, the model represents that a person has disease state d_u, for $0 \le u \le n$. Let p_u denote $P(ps_i\_df = cough \mid ps_i\_dx = d_u)$. We assume that p_u is distributed according to a Beta distribution, namely, $p_u \sim Beta(\alpha_u, \beta_u)$. We assessed the parameters of these Beta distributions based on real data and expert judgments.

2.2 The Safety-Net Model

The example safety-net model introduced in this paper is designed to detect diseases that have a probability of cough, call it p^*, that is not equal to $p_0, p_1, \ldots, p_n$. We will represent a distribution over p^*. We use d_- (d_+) to denote a hypothetical disease that has a distribution of the probability of cough that is a delta function at 0 (1). Correspondingly, $p_- = 0$ and $p_+ = 1$. Consider every pair of known diseases being modeled, d_u and d_v. We consider the possibility that p^* is between p_u and p_v, for every u and v, such that $u, v \in \{-, 0, 1, \cdots n, +\}$. Each such possibility, $p_u < p^* < p_v$, constitutes one instance of the d^* disease hypothesis, which we denote as d_{uv}.

For d_{uv} we define the distribution over p^* as follows. We assume that p^* is uniformly distributed between p_u and p_v, and that p_u and p_v are distributed as Beta(α_u, β_u) and Beta(α_v, β_v), respectively. We stochastically sample p^* according to these distributions to obtain a distribution over p^*.

2.3 Inference

We wish to derive $P(pop\_dx \mid data)$, where *data* denotes the status of the symptopm *cough* for every person in the population. We assume the status is either *cough* or *no cough* for people who come to the ED, and that it is always *unknown* for people who do not. We derive $P(pop\_dx \mid data)$ by deriving $P(data \mid pop\_dx)$, assessing $P(pop\_dx)$, and applying the Bayes rule.

We derive $P(data \mid pop\_dx)$ by setting pop_dx to be one of d_0, d_k or d^*, and then performing inference on the Bayesian network in Fig. 1. Inference is complicated by the fact that we have distributions over $P(ps_i\_df \mid ps_i\_dx)$, as described in Sections 2.1 and 2.2; thus, inference includes integrating over these distributions.

We applied a variation of the inference method given in [4], which is polynomial time in the number of people who come to the ED. For the detailed description of inference, please see [5].

3 Preliminary Evaluation

For simplicity, we will assume that the *fraction* node has the value 0.0001 with probability 1; this assumption is not necessary, although it does reduce computational complexity.

3.1 Creating the Datasets

We created 20 datasets (scenarios), assuming a population size of 100,000 people. Each scenario represents data on the population for one given day of interest. In the remainder of this section, we describe how we created a scenario.

We sampled a Poisson distribution with mean $\lambda = 90$ to determine the number of people who came to the ED on the given day without any outbreak disease. For each of these people, we sampled their cough status using the distributions defined in Section 2.1.

When simulating the presence of outbreak disease d_k in the population, we assumed that the value of the node *fraction* in Fig. 1 is 0.0001. We assumed 100,000 $\times$ 0.0001 = 10 people had d_k and came to the ED. For each of these 10 people, we sampled from the distribution of $P(ps_{i\_}df = cough \mid ps_{i\_}dx = d_k)$ to determine their cough status. We then combined these 10 cases with the simulated ED cases without outbreak disease in order to create a complete dataset for the scenario.

3.2 Experimental Setup

Let d_u and d_v be any two of the eight CDC Category A diseases [6], $d_u \neq d_v$. Table 1 shows the experiments for one pair of d_u and d_v, where d_u is the simulated outbreak disease. In experiment B1 (B2), there is an explicit modeling of disease d_u (d_v), while in experiment A1 and A2 we also include safety-net disease d^* in the model. In each of the four experiments, we compute the likelihood ratio (*LR*) $P(data \mid outbreak)$ / $P(data \mid non\text{-}outbreak)$ as given by Eq. 1. In experiment A1 (A2), the sum in Eq. 1 is taken over d_u (d_v) *and* d^*; in contrast, in experiment B1 (B2) the sum of pop_dx consists only of the term d_u (d_v). For any given outbreak disease d_k being modeled, we assumed that $P(pop\_dx = d_k \mid outbreak) = P(pop\_dx = d^* \mid outbreak) = 0.5$.

Table 1. A 2 × 2 table that summarizes the experiments

	A	B
1	Model d_0, d_u, d^*. Simulate outbreak cases from d_u.	Model d_0, d_u. Simulate outbreak cases from d_u.
2	Model d_0, d_v, d^*. Simulate outbreak cases from d_u.	Model d_0, d_v. Simulate outbreak cases from d_u.

To investigate the degree to which modeling the safety-net disease d^* has an impact on detection performance, we made d_u and d_v to be all possible pairs of the eight outbreak diseases and carried out 8 × 7 = 56 sets of experiments. In each set, we computed the mean RR_1 and the mean RR_2 over the 20 scenarios, where for a given scenario $RR_1 = LR(S_{B1}) / LR(S_{A1})$ and $RR_2 = LR(S_{A2}) / LR(S_{B2})$, where $S_{A1} = \{d_u, d^*\}$, $S_{B1} = \{d_u\}$, $S_{A2} = \{d_v, d^*\}$, and $S_{B2} = \{d_v\}$. Our hypothesis is that usually $RR_2 > RR_1$, which supports that modeling d^* is doing more good than harm in detecting outbreak diseases.

$$LR = \frac{\sum_{pop\_dx} P(data \mid pop\_dx)P(pop\_dx \mid outbreak)}{P(data \mid pop\_dx = d_0)} \tag{1}$$

4 Results and Discussion

Recall that the distribution of $P(ps_i\_df = cough \mid ps_i\_dx = d_k)$ was assessed by a domain expert, for $1 \le k \le 8$. We sorted the eight outbreak diseases by their expectations for $P(ps_i\_df = cough \mid ps_i\_dx = d_k)$ in ascending order. In Tables 2, 3 and 4, each row represents a disease d_u and each column represents a disease d_v. We list the eight diseases according to the sorted order from left to right and from up and down. The closer d_u and d_v are to the diagonal, the closer are their means, and in a sense the closer the two diseases are in their symptomatic presentation.

Table 2. The mean RR_2 given different d_u (rows) and d_v (columns)

d_v / d_u	small pox	cryptospo -ridiosis	early anthrax	late anthrax	asthma	influenza	early plague	late plague
small pox	-	0.84	0.84	1.94	2.88	2.33	4.65	6.74
cryptospo -ridiosis	0.90	-	0.84	1.66	2.27	1.98	3.42	4.62
early anthrax	1.05	0.97	-	1.40	1.77	1.64	2.48	3.15
late anthrax	2.13	1.90	1.38	-	0.91	0.93	1.02	1.11
asthma	2.54	2.29	1.59	0.82	-	0.84	0.87	0.91
influenza	2.75	2.54	1.72	0.82	0.82	-	0.88	0.92
early plague	2.76	2.51	1.70	0.80	0.78	0.81	-	0.84
late plague	3.35	3.14	2.04	0.77	0.73	0.75	0.75	-

Table 2 shows the mean RR_2 given different combinations of d_u and d_v. The mean RR_2 tends to increase from the diagonal to the top right and to the bottom left corners. It shows that when there is an unexpected disease d_u present, the greater the difference in presentation of d_v relative to d_u, the greater the expected benefit from modeling d^*.

Table 3 shows the mean RR_1. Since there is no disease d_v involved in deriving RR_1, every row has the same values. Table 3 shows that RR_1 is quite stable at a value only modestly greater than 1, which provides support that when there is no unanticipated disease present, modeling d^* only weakly degrades the detection of d_u.

Table 3. The mean RR_1 given different d_u

d_u \ d_v	small pox	cryptospo -ridiosis	early anthrax	late anthrax	asthma	influenza	early plague	late plague
small pox	-	1.21	1.21	1.21	1.21	1.21	1.21	1.21
cryptospo -ridiosis	1.18	-	1.18	1.18	1.18	1.18	1.18	1.18
early anthrax	1.14	1.14	-	1.14	1.14	1.14	1.14	1.14
late anthrax	1.15	1.15	1.15	-	1.15	1.15	1.15	1.15
asthma	1.24	1.24	1.24	1.24	-	1.24	1.24	1.24
influenza	1.21	1.21	1.21	1.21	1.21	-	1.21	1.21
early plague	1.23	1.23	1.23	1.23	1.23	1.23	-	1.23
early plague	1.34	1.34	1.34	1.34	1.34	1.34	1.34	-

We performed the sign tests to calculate P-values over the null hypothesis H_o: $RR_1 > RR_2$ versus the alternative hypothesis H_a: $RR_1 \leq RR_2$. Table 4 shows the P-values given different combinations of d_u and d_v. Notice that the P-values close to the diagonal are very big, so that we cannot reject the null hypothesis, while the P-values away from the diagonal are zeros, which rejects the null hypothesis. Table 4 provides support that modeling d^* helps detect unanticipated diseases more than it interferes with detecting known diseases.

Table 4. A table shows the P-values given different combinations of d_u and d_v

d_u \ d_v	small pox	cryptospo -ridiosis	early anthrax	late anthrax	asthma	influenza	early plague	late plague
small pox	-	0.99	0.99	0	0	0	0	0
cryptospo -ridiosis	0.99	-	1	0	0	0	0	0
early anthrax	0.99	0.99	-	0	0	0	0	0
late anthrax	0	0	0	-	0.99	0.99	0.99	0.99
asthma	0	0	0	1	-	1	1	1
influenza	0	0	0	1	1	-	0.99	0.98
early plague	0	0	0	1	1	1	-	1
late plague	0	0	0	1	1	1	1	-

5 Conclusion and Future Work

This paper introduced a Bayesian method for detecting disease outbreaks that combines a specific detection method with a non-specific method. Preliminary results provide support that this hybrid approach helps detect unexpected diseases more than it interferes with detecting known diseases.

We plan to test this approach on real datasets and evaluate its detection performance using other measures, such as AMOC curves [2].

Acknowledgments. This research was funded by a grant from the National Science Foundation (NSF IIS-0325581).

References

1. Wong, W.-K., *Data mining for early disease outbreak detection [Doctoral Dissertation].* 2004, Carnegie Mellon University: Pittsburgh.
2. Fawcett, T. and F. Provost, *Adaptive fraud detection.* Data Mining and Knowledge Discovery, 1997. **1**(3): p. 291-316.
3. Denning, D., *An intrusion-detection model.* IEEE Transactions on Software Engineering, 1987.
4. Cooper, G.F., *A Bayesian method for learning belief networks that contain hidden variables.* Journal of Intelligent Information Systems, 1995. **4**: p. 71-88.
5. Shen, Y. and G.F. Cooper, *Bayesian disease outbreak detection that includes a model of unknown disease.* Department of Medical Informatics, University of Pittsburgh: Technical Report No. DBMI-07-351, 2007.
6. Hutwagner, L., W. Thompson, and G.M. Seeman, *The bioterrorism preparedness and response early aberration reporting system (EARS).* Journal of Urban Health, 2003. **80** (2, Supplement 1): p. i89-i96.

Spatial Epidemic Patterns Recognition Using Computer Algebra

Doracely Hincapié[1] and Juan Ospina[2]

[1] Group of Epidemiology,
National School of Public Health
University of Antioquia
Medellin, Colombia
doracely@guajiros.udea.edu.co
[2] Logic and Computation Group
School of Sciences and Humanities
EAFIT University
Medellin, Colombia
judoan@une.net.co

Abstract. An exploration in Symbolic Computational bio-surveillance is showed. The main obtained results are that the geometry of the habitat determines the critical parameters via the zeroes of the Bessel functions and the explicit forms of the static and non-static spatial epidemic patterns.

Keywords: spatial epidemic patterns, pan-endemic state, pan-epidemic state, critical parameter, velocity of propagation, endemic boundary, special functions.

1 Introduction

A very interesting issue, which remains practically unexplored, it is concerned with the spatial propagation of the diseases inside habitats with boundaries and the patterns or spatial profiles that the disease displays. In the present work we consider the question about the spatial patterns that the infectious diseases generate, for the case of a circular habitat with endemic boundary, which is initially free of infection. The present work is an exploration in the land of the Symbolic Computational Spatial Bio-Surveillance (SCSBS).

2 Methods

We consider here a circular habitat with a radio a. The model is as follows [1-3]

$$\frac{\partial}{\partial t} X(r, t) = \frac{\eta \left(\left(\frac{\partial}{\partial r} X(r, t) \right) + r \left(\frac{\partial^2}{\partial r^2} X(r, t) \right) \right)}{r} - \beta \, X(r, t) \, Y(r, t) \tag{1}$$

D. Zeng et al. (Eds.): BioSurveillance 2007, LNCS 4506, pp. 216–221, 2007.

$$\frac{\partial}{\partial t} Y(r, t) = \frac{\eta\left(\left(\frac{\partial}{\partial r} Y(r, t)\right) + r\left(\frac{\partial^2}{\partial r^2} Y(r, t)\right)\right)}{r} + \beta\, X(r, t)\, Y(r, t) - \gamma\, Y(r, t) \quad (2)$$

$$\frac{\partial}{\partial t} Z(r, t) = \frac{\eta\left(\left(\frac{\partial}{\partial r} Z(r, t)\right) + r\left(\frac{\partial^2}{\partial r^2} Z(r, t)\right)\right)}{r} + \gamma\, Y(r, t) \quad (3)$$

Where $X(r,t)$ is the susceptible density, $Y(r,t)$ is the infective density, $Z(r,t)$ is the density of removed, β is the infectiousness, γ the recovery constant and η is the diffusivity.

2.1 Static Spatial Epidemic Patterns

The following patterns for infected and removed people are obtained if (6) is satisfied.

$$Y(r) = \frac{Y_b\, J_0\left(\sqrt{\frac{\beta\,N-\gamma}{\eta}}\, r\right)}{J_0\left(\sqrt{\frac{\beta\,N-\gamma}{\eta}}\, a\right)} \quad (4)$$

$$Z(r) = Z_b + \frac{\gamma\, Y_b\left(\frac{J_0\left(\sqrt{\frac{\beta\,N-\gamma}{\eta}}\, r\right)}{J_0\left(\sqrt{\frac{\beta\,N-\gamma}{\eta}}\, a\right)} - 1\right)}{\beta\,N-\gamma} \quad (5)$$

The total number of infected individuals is given by (7).

$$(1 < R_0) < 1 + \frac{5.784025\,\eta}{\gamma\,a^2} \quad (6)$$

$$Y = \frac{2\,\pi\, Y_b\, a\, \sqrt{\eta}\, J_1\left(\frac{\sqrt{\beta\,N-\gamma}\; a}{\sqrt{\eta}}\right)}{J_0\left(\sqrt{\frac{\beta\,N-\gamma}{\eta}}\, a\right)\sqrt{\beta\,N-\gamma}} \quad (7)$$

The Fig.1. illustrates the pan-endemic threshold condition (6).

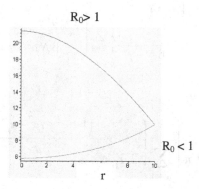

Fig. 1. Two static epidemic profiles of infected people according with the equation (4) are showed. This is an illustration of the pan-endemic threshold condition (6). The curve of the top corresponds to the super-critical case and the curve of the bottom corresponds to the sub-critical case. In the super- critical case the infection is more protuberant at the center of the habitat $r = 0$ but in the sub-critical case the infection remains confined to the endemic boundary $r=a$.

2.2 Dynamic Spatial Epidemic Patterns

The equation (8) can be derived [1], for the case of a circular habitat with endemic boundary when the infective profile Y(r,t) is restricted by the boundary condition Y(a,t)=Y_b and it is subjected to the initial condition Y(r,0)=0. The initial condition indicates that initially the habitat is free of infection but the boundary condition represents the case when the disease is permanently introduced from the endemic boundary. This boundary condition of the Dirichlet kind corresponds to an emergent infectious disease but a boundary condition of the Newman kind is a model of a biological attack. Here only the Dirichlet condition is studied.

$$Y(r,t) = \frac{Y_b J_0\left(\sqrt{\frac{\beta N - \gamma}{\eta}}\, r\right)}{J_0\left(\sqrt{\frac{\beta N - \gamma}{\eta}}\, a\right)} + \sum_{n=1}^{\infty}\left(\frac{2 Y_b J_0\left(\frac{\alpha_n r}{a}\right) e^{\left(\frac{\left(\beta N a^2 - \gamma a^2 - \alpha_n^2 \eta\right)t}{a^2}\right)}}{J_1(\alpha_n)\left(\beta N a^2 - \gamma a^2 - \alpha_n^2 \eta\right)} \alpha_n \eta\right) \tag{8}$$

The corresponding critical parameter is quantized as (9) [1] , where n is an positive integer such $n \geq 1$, and α_n are the zeros of the Bessel function J_0. The fundamental critical parameter corresponds to n=1 with α_1=2.405 and it is given like (10) [1].

$R_{0,n} = \dfrac{\beta N}{\gamma + \dfrac{\alpha_n^2 \eta}{a^2}}$ $\qquad$ (9)	$R_{0,1} = \dfrac{\beta N}{\gamma + \dfrac{5.784025\,\eta}{a^2}}$ $\qquad$ (10)

For hence the pan-epidemic state is established in the circular habitat with endemic boundary only when $R_{0,1} > 1$.

The total number of infectious individuals for every time in the habitat is computed as

$$Y(t) = 2\pi\left[\frac{Y_b\sqrt{\eta}\, a J_1\left(\sqrt{\frac{\beta N - \gamma}{\eta}}\, a\right)}{J_0\left(\sqrt{\frac{\beta N - \gamma}{\eta}}\, a\right)\sqrt{\beta N - \gamma}} + \sum_{n=1}^{\infty}\left(\frac{2 Y_b a^2 e^{\left(\frac{\left(\beta N a^2 - \gamma a^2 - \alpha_n^2 \eta\right)t}{a^2}\right)}}{\beta N a^2 - \gamma a^2 - \alpha_n^2 \eta}\eta\right)\right] \tag{11}$$

and the incidence is given by

$$\frac{d}{dt}Y(t) = 2\pi\left(\sum_{n=1}^{\infty}\left(2 Y_b e^{\left(\frac{\left(\beta N a^2 - \gamma a^2 - \alpha_n^2 \eta\right)t}{a^2}\right)}\eta\right)\right) \tag{12}$$

The velocity of propagation of the disease in the circular habitat with endemic boundary and when the initial focus of infection is located justly in such endemic boundary, can be defined by (13) where r(t) is the instantaneous center of mass for the profile of infectious people at time t, Y(r,t); and it is given by (14).

$$v(t) = -\left(\frac{d}{dt}r(t)\right) \tag{13}$$

$$r(t) = \frac{\displaystyle\int_0^a Y(r, t)\, r^2\, dr}{\displaystyle\int_0^a Y(r, t)\, r\, dr} \tag{14}$$

The Fig. 2 gives an illustration of the equation (8). The Fig. 3 is depicted according with the equations (13) and (14).

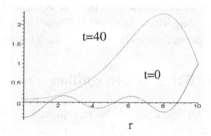

t=40

t=0

r

Fig. 2. Two spatial epidemic profiles of infected people for two different times, according with the equation (8) are showed. The curve at the top is for t=40 and the curve at the bottom is for t=0. Initially the infection is only concentrated at the endemic boundary but for t =40 the infection is yet present in the interior of the habitat and the more affected zone is not now the boundary.

2.3 Parameter Estimation

Using the previously given equations jointly with real data it is possible to estimate the basic parameters and the critical parameters. We use the least square method for the optimal adjustment between observed and predicted static and non-stationary spatial epidemic patterns. Assuming that the real data are organized as the pattern y(r,t) and the predicted pattern has the form Y(r,t, p_1, p_2, ...) where p_i are the parameters of the model, it is possible to construct the   summation of squares of deviations given by (15).

$$S(r, t, p_i) = \sum_j \int_0^a r^2\, (y(r, t_j) - Y(r, t_j, p_i))^2\, dr \tag{15}$$

where t_j are the sampling times. and for hence the optimal adjustment is obtained when the basic parameters of the model are determined as the solutions of the following system of equations.

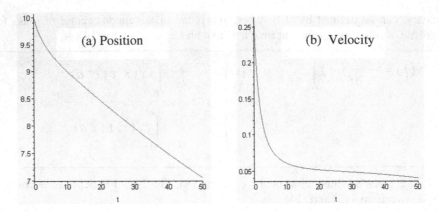

Fig. 3. (a) Curve for the trajectory of the infection according with (14). (b) Curve for the velocity of propagation according with (13) and (14). These curves were computed assuming that the fundamental pan-epidemic threshold condition $R_{0,1} > 1$ is satisfied.

3   Spatial Epidemic Patterns Recognition

Using the equation (15) for the case of the equations (4) and (8), it is possible to estimate the epidemic parameters, to obtain the critical parameters and to derive an approximation for the velocity of propagation.   Such substitutions generate very complicated integrals which are difficult to hand at general and they will not be presented here by space reasons. Only the calculations for the equation (4) will be considered here. Specifically, assuming that the observed profile $y(r)$ has the form $y(r) = AJ_0(\delta r)$, where A and δ are known parameters, the substitution of (4) in (15) gives an integral which can be symbolically computed using Maple and the final result is showed by the equation (16)

$$S(\beta, \gamma) = \frac{1}{3} a^3 A^2 \, F\!\left(\left[\frac{1}{2}, \frac{3}{2}\right], \left[1, 1, \frac{5}{2}\right], -a^2 \delta^2\right) - \int_0^a \frac{r^2 A J_0(\delta r) Y_b J_0\!\left(\sqrt{\frac{\beta N - \gamma}{\eta}}\, r\right)}{J_0\!\left(\sqrt{\frac{\beta N - \gamma}{\eta}}\, a\right)} \, dr$$

$$+ \frac{1}{3} \frac{a^3 Y_b^2 \, F\!\left(\left[\frac{1}{2}, \frac{3}{2}\right], \left[1, 1, \frac{5}{2}\right], -\frac{(\beta N - \gamma) a^2}{\eta}\right)}{J_0\!\left(\sqrt{\frac{\beta N - \gamma}{\eta}}\, a\right)^2}$$

(16)

where F is the hypergeometric function. The optimization of (16) gives the following results:

$A = \dfrac{Y_b}{J_0\left(\sqrt{\dfrac{\beta N - \gamma}{\eta}}\, a\right)}$	$\delta = \sqrt{\dfrac{\beta N - \gamma}{\eta}}$	$R_0 = 1 + \dfrac{\eta\,\delta^2}{\gamma}$
	$\beta = \dfrac{\gamma + \eta\,\delta^2}{N}$	$\delta < \dfrac{2.405}{a}$

4 Conclusions

1. The critical parameter for the existence of the pan-endemic configuration depends only over epidemical parameters but it is limited by the geometry of the habitat by mean of the first zero of J_0.
2. The critical parameter for the existence of the pan-epidemic configuration is quantized in a similar way to a certain systems of quantum mechanics and it depends not only over the usual epidemical parameters but that also it depends on the geometry via the zeros of J_0.

References

1. Hincapie, D., Ospina J.: Basic Reproductive Rate of a Spatial Epidemic Model using Computer Algebra Software. Proceedings of METMBS'05. Valafar & Valafar (eds). CSREA Press, 2005.
2. Guarin N., Ospina J.: Analytical Solution of a Spatial SIR epidemic model with memory inside a circular habitat with endemic memory using CAS. Proceedings of BIOCOMP'06. Arabnia & Valafar (eds). CSREA Press, 2006.
3. Juan Ospina and Doracelly Hincapié. Mackendrick: A Maple package oriented to symbolic computational epidemiology. Lecture Notes in Computer Science 2006; S 3991: 920–923

Detecting Conserved RNA Secondary Structures in Viral Genomes: The RADAR Approach

Mugdha Khaladkar and Jason T.L. Wang

Bioinformatics Program and Department of Computer Science
New Jersey Institute of Technology, Newark, NJ 07102, USA
jason.t.wang@njit.edu

Abstract. Conserved regions, or motifs, present among RNA secondary structures serve as a useful indicator for predicting the functionality of the RNA molecules. Automated detection or discovery of these conserved regions is emerging as an important research topic in health and disease informatics. In this short paper we present a new approach for detecting conserved regions in RNA secondary structures by the use of constrained alignment and apply the approach to finding structural motifs in some viral genomes. Our experimental results show that the proposed approach is capable of efficiently detecting conserved regions in the viral genomes and is comparable to existing methods. We implement our constrained structure alignment algorithm into a web server, called RADAR. This web server is fully operational and accessible on the Internet at http://datalab.njit.edu/biodata/rna/RSmatch/server.htm.

1 Introduction

RNA molecules play various roles in the cell [1-3]. Their functionality depends not only on the sequence information but to a large extent on their secondary structures. It would be more cost effective if one were able to determine RNA structure by computational means rather than by using biochemical methods. So, the development of computational predictive approaches of RNA structure is essential [4-7]. RNA structure prediction is usually based on the thermodynamics of RNA folding [8-10] or phylogenetic conservation of base-paired regions [11-14].

In this short paper, we present a methodology for the detection of conserved regions, or motifs, in RNA secondary structures. We adopt a comparative approach by carrying out RNA structure alignment. The alignment has been designed so as to be constrained based upon the properties of the RNA structure. Our constrained alignment algorithm is an upward extension of our previously developed RSmatch method for comparing RNA secondary structures, which did not consider constraints in the structures being aligned [7]. We have implemented the constrained structure alignment algorithm into a web server, called RADAR (acronym for *RNA Data Analysis and Research*) [15]. This web server provides multiple capabilities for RNA structure alignment data analysis, which includes pairwise structure alignment, multiple structure alignment, constrained structure alignment and consensus structure prediction. Our aim behind developing this web server is to develop a versatile tool that provides a computationally efficient platform for performing several tasks related to RNA

D. Zeng et al. (Eds.): BioSurveillance 2007, LNCS 4506, pp. 222–227, 2007.

secondary structure. RADAR has been developed using Perl-CGI and Java. The web server has a user friendly interface that is easy to understand even for novice users.

The rest of this short paper is organized as follows. Section 2 presents the constrained structure alignment algorithm. Section 3 reports implementation efforts, showing some features of the RADAR server, and describes experimental results. Section 4 concludes the paper and points out some directions for future research.

2 Constrained Structure Alignment

In practice, biologists favor integrating their knowledge about conserved regions into the alignment process to obtain biologically more meaningful similarity scores between RNAs. This motivates us to develop algorithms for constrained structure alignment (CSA). Our CSA method is developed by modifying the recurrence formulas of the dynamic programming algorithm used in RSmatch [7] to take into account the constraints, or conserved regions, in the structures being aligned. The time complexity of RSmatch is $O(mn)$, where m and n are the sizes of the two RNA structures respectively that are being aligned. The CSA method has the same time complexity. In practice, the CSA method can align two RNA structures in less than two seconds in wall clock time.

2.1 Method

The input of the CSA method is a query RNA structure and a database of RNA secondary structures. The method constructs the alignment between the query structure and each database structure based upon the knowledge of conserved regions in the query structure. The alignment score is dynamically varied so as to utilize the information of conserved regions. The alignment computed this way is able to detect structural similarity more accurately. The method comprises two main parts:

(I) Annotating a region in the query RNA structure as conserved
Each position of the conserved region in the query RNA structure is marked using a special character "*" underneath the position. This is termed as *binary conservation* since any position in the query RNA structure is treated to be either 100% conserved (if it is marked with "*") or not conserved at all. If it is found, from wet lab experiments or other sources, that a particular RNA structure contains a motif that we want to search for in other RNA structures in a database, then that particular RNA structure can be used as a query structure and that motif region can be marked to be conserved in the query structure. Under this circumstance, the user can adopt binary 0/1 conservation.

Another technique of applying constraints to structure alignment is using the concept of sequence logos [16]. First, the multiple sequence alignment of the RNA sequences under analysis is obtained using, for example, ClustalW (http://www.ebi.ac.uk/clustalw/). Then the frequency of each base at each position of an RNA sequence is calculated, denoted by $f(b,l)$. Using this information the uncertainty at each position is calculated by $H(l) = - \sum f(b,l) \, log_2 f(b,l)$, where $H(l)$ is the uncertainty at position l, b is one of the bases (A,C,G,U), and $f(b,l)$ is the frequency of base b at position l. The information content at each position is represented by the decrease in uncertainty as the binding site is located (or aligned), i.e. $R_{sequence}(l) = 2 - H(l)$, where $R_{sequence}(l)$ is the amount of information present at position l and 2 is the maximum uncertainty at any given position. Thus, the information content is a value between 0 and 2. We scale down this value to get a value between 0 and 1 which is then converted to the percentage conservation at each position of the RNA sequence and its secondary structure.

(II) Utilization of information about the conserved region
Two cases occur as we compute the alignment score between the query structure and a database structure where the query structure contains marked conserved regions.

1. *Comparison between non-conserved regions:* In this case the score assigned is the regular score that is derived from the scoring matrix used by RSmatch.
2. *Comparison involving conserved regions:* Here, we multiply the score obtained from the scoring matrix used by RSmatch by a factor λ that will cause the score to either increase or decrease by the λ value. This factor λ is determined by the type of conservation as discussed in more detail in the subsection on "Scoring Scheme".

2.2 Scoring Scheme

The factor by which the score should get magnified or diminished to take into account the conserved region is decided based upon the following: (i) the length of the conserved region; (ii) the length of the whole RNA sequence; (iii) the type of conservation that has been indicated; and (iv) any special conditions/preferences decided by the user.

In the default scenario, where knowledge about conservation is not used, the score is directly taken from the scoring matrix employed in RSmatch. For the binary conservation case, the default value for the factor λ is $\lambda = 2 - L/N$ where L is the length of the conserved region and N is the length of the whole RNA sequence. This ratio is then subtracted from a constant value (2, arbitrarily chosen) so that the bonus/penalty is inversely proportional to the length of the conserved region. If the conservation information is spread over 0-100%, as described earlier, these percentage values are passed along with the query RNA structure to the scoring engine and the alignment score varies based on these values.

3   Implementation and Experimental Results

We have implemented the proposed constrained structure alignment algorithm into the RADAR server. This web server together with a standalone downloadable version is freely available at http://datalab.njit.edu/biodata/rna/RSmatch/server.htm. RADAR accepts, as input data, either RNA sequences in the standard FASTA format or RNA secondary structures represented by the Vienna style Dot Bracket format [8]. For performing the constrained structure alignment between a query RNA structure and a database structure, we require the user to annotate the query RNA structure to indicate

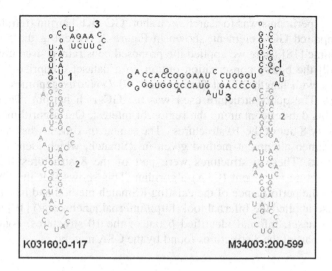

Fig. 1. An example showing the common region of two RNA secondary structures where the local matches in the two structures are highlighted with the green color

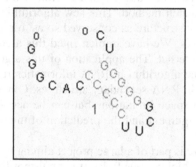

Fig. 2. The structure of a GC–rich hairpin that has been found to be conserved in the *Leviviridae* family [18]

which region is conserved by marking the region with "*". Thus, the input consists of an annotated query RNA structure and a database of RNA structures against which the constrained alignment is carried out. Upon completion of a constrained alignment job, RADAR presents the alignment results on a web page. In Figure 1, the common region of two RNA secondary structures given in an alignment result is portrayed using RnaViz [17], where the local matches in the two structures are highlighted with the green color.

We have conducted several experiments to evaluate the performance of the proposed constrained structure alignment algorithm by applying this method to finding structural motifs in viral genomes. Study of viral genomes has shown that they often contain functionally active RNA structural motifs that play an important role in the different stages of the life cycle of the virus [18]. Detection of such motifs or conserved regions would greatly assist the study of these viruses.

One of our experiments was to search for a short GC–rich hairpin (tetraloop) which follows an unpaired GGG element, shown in Figure 2, present at the 5′ end of the *Levivirus* genome [18]. Here we applied the proposed constrained structure alignment algorithm, with the binary conservation option, to a dataset comprising 6838 RNA structures each with length 200 nt formed from 10 *Levivirus* genomes and 4 other viral genomes. The query structure used was the GC–rich hairpin. There were 10 structures in this dataset containing the region of interest. Our algorithm was able to correctly identify 8 out of the 10 structures. The same experiment was repeated using the non-constrained alignment method given in RSmatch, which identified 6 out of the 10 structures. These 6 structures were part of the 8 structures found by the constrained structure alignment (CSA) algorithm. This shows that the CSA method improves upon the performance of the existing RSmatch method and has a better sensitivity. We also applied the Infernal tool (http://infernal.janelia.org/) [19] to this same viral genome dataset. Infernal identified 6 out of the 10 structures. Again, these 6 structures were part of the 8 structures found by the CSA method.

4   Conclusion

The proposed constrained structure alignment algorithm is an upward extension of our previously developed RSmatch method. This new algorithm allows the user to annotate some region in a query structure as conserved so as to produce biologically more meaningful similarity scores. We have implemented this algorithm into the RADAR server accessible on the Internet. The application of this algorithm to viral genomes demonstrates the use of the algorithm in RNA informatics research and its ability to detect conserved regions in RNA secondary structures. Other functions of RADAR, not described here due to space limitations but can be accessed on the Internet, include multiple structure alignment and the prediction of the consensus structure for a set of RNA sequences.

The work presented here is part of a large project aiming to build a cyber infrastructure (http://datalab.njit.edu/bioinfo) for RNA structural motif discovery in human, virus and trypanosome mRNAs. Human immunodeficiency virus type 1 is the causative agent of AIDS and is related to many cancers in humans. Hepatitis C virus is related to hepatocellular cancer in humans. *Trypanosoma brucei* causes African trypanosomaiasis, or sleeping sickness, in humans and animals in Africa. RNA motifs or conserved structures have been shown to play various roles in post-transcriptional control including mRNA translation, mRNA stability, and gene regulation, among others. This cyber infrastructure will contribute to integrated genomic and epidemiological data analysis, by enabling access, retrieval, comparison, analysis, and discovery of biologically significant RNA motifs through the Internet as well as the integration of these motifs with online biomedical ontologies.

Acknowledgements

We thank the anonymous reviewers whose comments help to improve the presentation and quality of this paper.

References

1. Wang, J.T.L., Zaki, M.J., Toivonen, H.T.T. and Shasha, D., eds. (2005) *Data Mining in Bioinformatics*. Springer, New York.
2. Wang, J.T.L., Wu, C.H. and Wang, P.P., eds. (2003) *Computational Biology and Genome Informatics*. World Scientific, Singapore.
3. Wang, J.T.L., Shapiro, B.A. and Shasha, D., eds. (1999) *Pattern Discovery in Biomolecular Data: Tools, Techniques and Applications*. Oxford University Press, New York.
4. Wang, J.T.L., Shapiro, B.A., Shasha, D., Zhang, K. and Chang, C.-Y. (1996) Automated discovery of active motifs in multiple RNA secondary structures. In *Proceedings of the 2nd International Conference on Knowledge Discovery and Data Mining*, pp. 70-75.
5. Wang, J.T.L. and Wu, X. (2006) Kernel design for RNA classification using support vector machines. *International Journal of Data Mining and Bioinformatics*, 1, 57-76.
6. Bindewald, E. and Shapiro, B.A. (2006) RNA secondary structure prediction from sequence alignments using a network of k-nearest neighbor classifiers. *RNA*, 12, 342-352.
7. Liu, J., Wang, J.T.L., Hu, J. and Tian, B. (2005) A method for aligning RNA secondary structures and its application to RNA motif detection. *BMC Bioinformatics*, 6, 89.
8. Hofacker, I.L. (2003) RNA secondary structure server. *Nucleic Acids Res.*, 31, 3429-3431.
9. Schuster, P., Fontana, W., Stadler, P.F. and Hofacker, I.L. (1994) From sequences to shapes and back: a case study in RNA secondary structures. *Proc. Roy. Soc. (London) B*, 255, 279-284.
10. Zuker, M. (1989) Computer prediction of RNA structure. *Methods Enzymol.*, 180, 262-288.
11. Akmaev, V.R., Kelley, S.T. and Stormo, G.D. (2000) Phylogenetically enhanced statistical tools for RNA structure prediction. *Bioinformatics*, 16, 501-512.
12. Gulko, B. and Haussler, D. (1996) Using multiple alignments and phylogenetic trees to detect RNA secondary structure. In *Proceedings of the 1st Pacific Symposium on Biocomputing*, pp. 350-367.
13. Knudsen, B. and Hein J. (2003) Pfold: RNA secondary structure prediction using stochastic context-free grammars. *Nucleic Acids Res.*, 31, 3423-3428.
14. Rivas, E. and Eddy, S.R. (1999) A dynamic programming algorithm for RNA structure prediction including pseudoknots. *J. Mol. Biol.*, 285, 2053-2068.
15. Khaladkar, M., Bellofatto, V., Wang, J.T.L., Tian, B. and Zhang, K. (2006) RADAR: an interactive web-based toolkit for RNA data analysis and research. In *Proceedings of the 6th IEEE Symposium on Bioinformatics and Bioengineering*, pp. 209-212.
16. Schneider, T.D. and Stephens, R.M. (1990) Sequence logos: a new way to display consensus sequences. *Nucleic Acids Res.*, 18, 6097-6100.
17. Rijk, P.D., Wuyts, J. and Wachter, R.D. (2003) RnaViz2: an improved representation of RNA secondary structure. *Bioinformatics*, 19, 299-300.
18. Hofacker, I.L., Stadler, P.F. and Stocsits, R.R. (2004) Conserved RNA secondary structures in viral genomes: a survey. *Bioinformatics*, 20, 149.
19. Eddy, S.R. (2002) A memory-efficient dynamic programming algorithm for optimal alignment of a sequence to an RNA secondary structure. *BMC Bioinformatics*, 3, 18.

Gemina: A Web-Based Epidemiology and Genomic Metadata System Designed to Identify Infectious Agents

Lynn M. Schriml[1], Aaron Gussman[1], Kathy Phillippy[1], Sam Angiuoli[1], Kumar Hari[2], Alan Goates[2], Ravi Jain[3], Tanja Davidsen[1], Anu Ganapathy[1], Elodie Ghedin[4], Steven Salzberg[1,5], Owen White[1], and Neil Hall[1]

[1] The Institute for Genomic Research, a Division of J. Craig Venter Institute
[2] Ibis Biosciences, a Division of Isis Pharmaceuticals, Inc.
[3] cBIO, Inc.
[4] University of Pittsburgh School of Medicine, Division of Infectious Diseases
[5] Center for Bioinformatics and Computational Biology, University of Maryland

The Gemina system (http://gemina.tigr.org) developed at TIGR is a tool for identification of microbial and viral pathogens and their associated genomic sequences based on the associated epidemiological data. Gemina has been designed as a tool to identify epidemiological factors of disease incidence and to support the design of DNA-based diagnostics such as the development of DNA signature-based assays. The Gemina database contains the full complement of microbial and viral pathogens enumerated in the Microbial Rosetta Stone database (MRS) [1]. Initially, curation efforts in Gemina have focused on the NIAID category A, B, and C priority pathogens [2] identified to the level of strains. For the bacterial NIAID category A-C pathogens, for example, we have included 38 species and 769 strains in Gemina. Representative genomic sequences are selected for each pathogen from NCBI's GenBank by a three tiered filtering system and incorporated into TIGR's Panda DNA sequence database. A single representative sequence is selected for each pathogen firstly from complete genome sequences (Tier 1), secondly from whole genome shotgun (WGS) data from genome projects (Tier 2), or thirdly from genomic nucleotide sequences from genome projects (Tier3). The list of selected accessions is transferred to Insignia when new pathogens are added to Gemina, allowing Insignia's Signature Pipeline [3] to be run for each pathogen identified in a Gemina query.

The distinct relationship between a pathogen and host for a particular infectious disease can be described by the concept of a "Chain of Infection". In Gemina, we have called this relationship an Infection System and have defined this concept by five components: Pathogen, Host, Disease, Transmission Method and Anatomy. The pathogen is described to the level of strain where possible using NCBI's taxonomy. The host includes both humans and animals and is also defined using NCBI's taxonomy. Disease data types includes the primary infectious disease caused by the pathogen as reported in the literature. The transmission method includes the direct or indirect method by which the pathogen infected the host or by which the pathogen moved to or from it's reservoir to a host. The anatomy includes the portal of entry where the pathogen enters the host. From searches of the public literature available in PubMed, we created lists of all combinations of the components of an Infection System to uniquely identify a disease. During curation we also collected geographic location, reservoir and symptom data for each pathogen and Infection System,

D. Zeng et al. (Eds.): BioSurveillance 2007, LNCS 4506, pp. 228–229, 2007.
© Springer-Verlag Berlin Heidelberg 2007

therefore, each Infection System has additional epidemiological information attached to it. To date, we have curated Infection Systems for all NIAID category A-C bacterial species and strains and all Influenza subtypes and strains. This data is available for querying in Gemina. In subsequent data releases we will also provide additional epidemiological data types: such as host gender, host age, collection date, PubMed IDs, links to additional web resources for each pathogens and Google Maps links for the geographic location data. The vast amount of biological information deposited in the public domain in the current era of genome-scale biology presents unique issues of information integration. To deal with this in Gemina we are utilizing ontologies and controlled vocabularies to regularize the epidemiological terms in the Gemina database, to provide hierarchical queries and data displays in the Gemina web interface and to provide consistent data retrieval [4]. The Gemina ontologies and controlled vocabularies for disease, symptom, anatomy, reservoir, geographic location and transmission method are available on TIGR's ftp site at ftp://ftp.tigr.org/pub/data/gemina. The Gemina system is built upon a relational database anchored by taxonomy identifiers from NCBI [5] for the pathogens and hosts. Gemina's web interface allows users to submit queries comprised of one or more epidemiology metadata types. This process allows a user to be highly selective about which strains to include in the signature design, such as only including virulent strains or strains that infect specific host species. Each query returns the associated Infection System rows with their curated pathogen, host, disease, transmission method and portal of entry (anatomy). The result for each Gemina query is linked to the representative genomic sequence for each pathogen. These results are displayed on Gemina's results page. This page includes additional tools including a link to the University of Maryland's Insignia Signature Pipeline.

Acknowledgements

This work was supported in part by DHS award W81XWH-05-2-0051.

References

1. Ecker, D.J., Sampath, R., Willett, P., Wyatt, J.R., Samant, V., Massire, C., Hall, T.A., Hari, K., McNeil, J.A., Buchen-Osmond, C., Budowle, B.: The Microbial Rosetta Stone Database: a compilation of global and emerging infectious microorganisms and bioterrorist threat agents. BMC Microbiol. 5 (2005) 19
2. NIAID http://www3.niaid.nih.gov/biodefense/PDF/cat.pdf
3. Insignia http://insignia.cbcb.umd.edu/
4. Smith B., Ceusters, W., Klagges, B., Kohler, J., Kumar, A., Lomax, J., Mungall, C., Neuhaus, F., Rector, A.L., Rosse, C. Relations in biomedical ontologies. Genome Biol. (2005) R46
5. National Center for Biotechnology Information (NCBI) Taxonomy http://www.ncbi.nlm.nih.gov/Taxonomy/taxonomyhome.html/

Internet APRS Data Utilization for Biosurveillance Applications

Tanya Deller[1], Rochelle Black[2], Francess Uzowulu[1], Vernell Mitchell[2],
and William Seffens[1]

[1] Biology Department,
[2] Computer and Information Science Department, Clark Atlanta University,
Atlanta, GA 30314
Wseffens@CAU.EDU

APRS is an abbreviation for Automatic Packet Reporting System, and is a method of broadcasting GPS positioning information in real time from packet radio-equipped stations. It was designed in the early 90's, but it has seen growth in the last few years due to user-friendly software such as WinAPRS or UI-View, and Kenwood's APRS enabled radio transceivers becoming available. APRS equipped stations send latitude and longitude information, as well as course, speed and altitude of mobile stations. These are commonly set up for use as Search and Rescue operations or special public events such as parades for tactical overviews. Even the International Space Station and a number of low-earth orbiting satellites have an APRS repeater on board, with amateur earth stations watching positions on their PC screens. Many stations also transmit weather data, which is collected for use by the US Weather Service.

In its most widely used form APRS is transported over the air using the AX.25 protocol at 1200 baud audio frequency-shift keying on frequencies located in the amateur 2-meter band (144.390 Mhz in the US). An extensive digital repeater, or digipeater network provides transport for APRS packets on these frequencies. Internet gateway stations (i-Gates) connect the on-air APRS network to the APRS Internet System (APRS-IS), which serves as a worldwide, high-bandwidth backbone for APRS data. Stations can tap into this stream directly. Databases connected to the APRS-IS allow web-based access to the data as well as more advanced data mining capabilities. The average bandwidth required for a single full APRS-IS feed is over 10,000 bits per second.

We are constructing a biosurveillance system that monitors the health of a large number of anonymous individuals from a group of volunteers transmitting this data as APRS packets. Collecting this data allows tracking health symptoms by GPS location and collected data will be used to correlate with CDC flu data in a program called Citizen Health Observer Program (CHOP). Participants would list some number of anonymous subjects for which the participant would have daily knowledge of any health symptoms. Every day, these participants would submit current health conditions of each person as packets transmitted to APRS-IS. Symptoms would be coded in terms of observables such as fever, cough, vomiting, etc., that would be relevant for bioweapons attack or flu monitoring. The added value of this collected data is that it is presented before the subjects seek medical attention. Patterns in the reporting of this data can be analyzed for correlation to collected CDC flu or other data. If 3,000 CHOP participants report on the health status of 10-20 subjects each, then close to 50,000 health reports would be processed daily. Since licensed radio

D. Zeng et al. (Eds.): BioSurveillance 2007, LNCS 4506, pp. 230–231, 2007.

operators are typically older, the participants in CHOP will sample a population more sensitive to flu or bioweapons outbreaks.

In addition to CHOP described above, use of collected APRS data could be used for detection of changes in transportation patterns to detect incidence of illness. APRS contains GPS information that could be used to identify locations of destinations for mobile stations. Patterns can be monitored for unusual trips to drug or grocery stores at unusual times that may be indicative of illness onset for a family member. Again, this data is collected several days before the subjects seek medical attention. This could give early warning of an impending disease outbreak. To test if this could be accomplished, we are writing a program to test the predictive capability of the APRS data. A Java program is being developed to implement Object Prediction for Traffic-cam Image Capture (OPTIC). This will monitor the APRS-IS data stream for identifiers (licensed radio amateur call signs) involved in this study. For example, KG4EYO-14 is the call sign of a GPS tracker installed in one of the author's cars, while KG4EYO-1 is a portable tracker that one of the other authors will temporally install in their vehicle. The packets for theses call signs are parsed for GPS location, direction of travel and speed. A table of Internet accessible traffic cameras in the Atlanta (GA) area lists the web address of the photo images, the GPS location of the camera, and the camera's update frequency. The nearest traffic camera in the table to the predicted track of the target vehicle is identified and a predicted time at which the vehicle is in view of the traffic camera is set. If no other APRS-IS data packets are received for the target vehicle, three web traffic camera image captures are executed based on the camera's update frequency. Accuracy of target vehicle predicted locations would be assessed as the OPTIC program is modified to include algorithms with greater knowledge of roads and traffic. Successful capture of target vehicles in the traffic camera images would indicate we could accurately predict if these vehicles are arriving at drug or grocery stores from APRS-IS data. The OPTIC program's table of traffic cameras will be increased to contain the GPS locations of drug stores, grocery stores, and health clinics in the Atlanta area. Pattern detection routines will be added to data mine the APRS-IS stream for interruption of daily vehicle patterns. Interruptions by trips to health-related vehicle stops will be correlated to personal logs kept by the authors to score when the event is due to an illness. The pattern detection routines can then be adjusted to give greater predictive measure for subject illness by vehicle traffic pattern analysis.

In summary, APRS data contains real time GPS, email content, weather and other telemetry data that is easily captured from Internet sites. This information can be data mined for biosurveillance applications without normal privacy concerns due to FCC regulations of Part 97 communications. Several years of archived APRS data is available from the corresponding author. This data provides a test bed for algorithm development in biosurveillance research.

Acknowledgments. This work was supported by NIH/NIGMS/MBRS/ SCORE/RISE (SCORE grant #S06GM08247).

Author Index

Allegra, John R.   196
Angiuoli, Sam   228

Bauer, David   71
Bettencourt, Luís M.A.   79
Black, Rochelle   230
Bloom, Steve   159
Brammer, Lynnette   159
Bresee, Joseph S.   159
Brown, Philip   196
Buckeridge, David L.   190
Burgess, Colleen R.   180
Burkom, Howard   59, 103

Carmeli, Boaz   147
Castillo-Chavez, Carlos   79
Chen, Hsinchun   11, 23, 134, 169
Chen, Yi-Da   23
Chowell, Gerardo   79
Cochrane, Dennis   196
Cohen, Steven A.   47
Cooper, Gregory F.   209
Copeland, John   159

Davidsen, Tanja   228
Deller, Tanya   230
Doshi, Meena   114
Dubrawski, Artur   124

English, Roseanne   159
Eshel, Tzilla   147

Fefferman, Nina   114
Ford, Daniel   147
Funk, Julie   1

Ganapathy, Anu   228
Ghedin, Elodie   228
Goates, Alan   228
Goodall, Colin R.   196
Greenshpan, Ohad   147
Gussman, Aaron   228

Halász, Sylvia   196
Hall, Neil   228
Hari, Kumar   228

Hicks, Peter   159
Higgs, Brandon W.   71
Hincapié, Doracelly   216
Hsiao, Jin-Yi   11
Hu, Paul Jen-Hwa   134

Jain, Ravi   228

Kaufman, James   147
Khaladkar, Mugdha   222
Kikuchi, Kiyoshi   202
King, Chwan-Chuen   11, 23
Knoop, Sarah   147

Lant, Timothy   79
Larson, Catherine A.   134
Lent, Arnold   196
Lofgren, Eric   114
Lu, Hsin-Min   11

McMurray, Paul   159
Mitchell, Vernell   230
Mohtashemi, Mojdeh   71
Murphy, Sean   59, 103

Naumova, Elena N.   47, 114
Nuño, Miriam   37

Ohkusa, Yasushi   202
Okabe, Nobuhiko   202
Ospina, Juan   216

Pagano, Marcello   37
Perez, Andrés   169
Phillippy, Kathy   228
Podgornik, Michelle N.   159
Postema, Alicia   159

Rajala-Schultz, Päivi   1
Ram, Roni   147
Renly, Sondra   147
Ribeiro, Ruy M.   79
Roure, Josep   124

Salzberg, Steven   228
Saville, William   1

Schneider, Jeff   124
Schriml, Lynn M.   228
Seffens, William   230
Shaffer, Loren   1
Shen, Yanna   209
Shih, Fuh-Yuan   11
Shmueli, Galit   91
Sugawara, Tamie   202

Tamblyn, Robyn   190
Taniguchi, Kiyosu   202
Thompson, William W.   159
Thurmond, Mark   169
Tokars, Jerome I.   159
Tseng, Chunju   23, 134, 169

Uhde, Kristin B.   159
Uzowulu, Francess   230

Verma, Aman   190

Wagner, Michael   1
Wallstrom, Garrick   1
Wang, Jason T.L.   222
White, Owen   228
Wittum, Thomas   1
Wu, Tsung-Shu Joseph   11, 23

Yahav, Inbal   91

Zeng, Daniel   11, 134, 169

Lecture Notes in Computer Science

For information about Vols. 1–4374

please contact your bookseller or Springer

Vol. 4510: P. Van Hentenryck, L. Wolsey (Eds.), Integration of AI and OR Techniques in Constraint Programming for Combinatorial Optimization Problems. X, 391 pages. 2007.

Vol. 4506: D. Zeng, I. Gotham, K. Komatsu, C. Lynch, M. Thurmond, D. Madigan, B. Lober, J. Kvach, H. Chen (Eds.), Intelligence and Security Informatics: Biosurveillance. XI, 234 pages. 2007.

Vol. 4486: M. Bernardo, J. Hillston (Eds.), Formal Methods for Performance Evaluation. VII, 469 pages. 2007.

Vol. 4483: C. Baral, G. Brewka, J. Schlipf (Eds.), Logic Programming and Nonmonotonic Reasoning. IX, 327 pages. 2007. (Sublibrary LNAI).

Vol. 4482: A. An, J. Stefanowski, S. Ramanna, C.J. Butz, W. Pedrycz, G. Wang (Eds.), Rough Sets, Fuzzy Sets, Data Mining and Granular Computing. XIV, 585 pages. 2007. (Sublibrary LNAI).

Vol. 4481: J.T. Yao, P. Lingras, W.-Z. Wu, M. Szczuka, N.J. Cercone, D. Ślęzak (Eds.), Rough Sets and Knowledge Technology. XIV, 576 pages. 2007. (Sublibrary LNAI).

Vol. 4480: A. LaMarca, M. Langheinrich, K.N. Truong (Eds.), Pervasive Computing. XIII, 369 pages. 2007.

Vol. 4472: M. Haindl, J. Kittler, F. Roli (Eds.), Multiple Classifier Systems. XI, 524 pages. 2007.

Vol. 4471: P. Cesar, K. Chorianopoulos, J.F. Jensen (Eds.), Interactive TV: a Shared Experience. XIII, 236 pages. 2007.

Vol. 4470: Q. Wang, D. Pfahl, D.M. Raffo (Eds.), Software Process Change – Meeting the Challenge. XI, 346 pages. 2007.

Vol. 4464: E. Dawson, D.S. Wong (Eds.), Information Security Practice and Experience. XIII, 361 pages. 2007.

Vol. 4463: I. Măndoiu, A. Zelikovsky (Eds.), Bioinformatics Research and Applications. XV, 653 pages. 2007. (Sublibrary LNBI).

Vol. 4462: D. Sauveron, K. Markantonakis, A. Bilas, J.-J. Quisquater (Eds.), Information Security Theory and Practices. XII, 255 pages. 2007.

Vol. 4459: C. Cérin, K.-C. Li (Eds.), Advances in Grid and Pervasive Computing. XVI, 759 pages. 2007.

Vol. 4453: T. Speed, H. Huang (Eds.), Research in Computational Molecular Biology. XVI, 550 pages. 2007. (Sublibrary LNBI).

Vol. 4452: M. Fasli, O. Shehory (Eds.), Agent-Mediated Electronic Commerce. VIII, 249 pages. 2007. (Sublibrary LNAI).

Vol. 4450: T. Okamoto, X. Wang (Eds.), Public Key Cryptography – PKC 2007. XIII, 491 pages. 2007.

Vol. 4448: M. Giacobini et al. (Ed.), Applications of Evolutionary Computing. XXIII, 755 pages. 2007.

Vol. 4447: E. Marchiori, J.H. Moore, J.C. Rajapakse (Eds.), Evolutionary Computation,Machine Learning and Data Mining in Bioinformatics. XI, 302 pages. 2007.

Vol. 4446: C. Cotta, J. van Hemert (Eds.), Evolutionary Computation in Combinatorial Optimization. XII, 241 pages. 2007.

Vol. 4445: M. Ebner, M. O'Neill, A. Ekárt, L. Vanneschi, A.I. Esparcia-Alcázar (Eds.), Genetic Programming. XI, 382 pages. 2007.

Vol. 4444: T. Reps, M. Sagiv, J. Bauer (Eds.), Program Analysis and Compilation, Theory and Practice. X, 361 pages. 2007.

Vol. 4443: R. Kotagiri, P.R. Krishna, M. Mohania, E. Nantajeewarawat (Eds.), Advances in Databases: Concepts, Systems and Applications. XXI, 1126 pages. 2007.

Vol. 4440: B. Liblit, Cooperative Bug Isolation. XV, 101 pages. 2007.

Vol. 4439: W. Abramowicz (Ed.), Business Information Systems. XV, 654 pages. 2007.

Vol. 4438: L. Maicher, A. Sigel, L.M. Garshol (Eds.), Leveraging the Semantics of Topic Maps. X, 257 pages. 2007. (Sublibrary LNAI).

Vol. 4433: E. Şahin, W.M. Spears, A.F.T. Winfield (Eds.), Swarm Robotics. XII, 221 pages. 2007.

Vol. 4432: B. Beliczynski, A. Dzielinski, M. Iwanowski, B. Ribeiro (Eds.), Adaptive and Natural Computing Algorithms, Part II. XXVI, 761 pages. 2007.

Vol. 4431: B. Beliczynski, A. Dzielinski, M. Iwanowski, B. Ribeiro (Eds.), Adaptive and Natural Computing Algorithms, Part I. XXV, 851 pages. 2007.

Vol. 4430: C.C. Yang, D. Zeng, M. Chau, K. Chang, Q. Yang, X. Cheng, J. Wang, F.-Y. Wang, H. Chen (Eds.), Intelligence and Security Informatics. XII, 330 pages. 2007.

Vol. 4429: R. Lu, J.H. Siekmann, C. Ullrich (Eds.), Cognitive Systems. X, 161 pages. 2007. (Sublibrary LNAI).

Vol. 4427: S. Uhlig, K. Papagiannaki, O. Bonaventure (Eds.), Passive and Active Network Measurement. XI, 274 pages. 2007.

Vol. 4426: Z.-H. Zhou, H. Li, Q. Yang (Eds.), Advances in Knowledge Discovery and Data Mining. XXV, 1161 pages. 2007. (Sublibrary LNAI).

Vol. 4425: G. Amati, C. Carpineto, G. Romano (Eds.), Advances in Information Retrieval. XIX, 759 pages. 2007.

Vol. 4424: O. Grumberg, M. Huth (Eds.), Tools and Algorithms for the Construction and Analysis of Systems. XX, 738 pages. 2007.

Vol. 4423: H. Seidl (Ed.), Foundations of Software Science and Computational Structures. XVI, 379 pages. 2007.

Vol. 4422: M.B. Dwyer, A. Lopes (Eds.), Fundamental Approaches to Software Engineering. XV, 440 pages. 2007.

Vol. 4421: R. De Nicola (Ed.), Programming Languages and Systems. XVII, 538 pages. 2007.

Vol. 4420: S. Krishnamurthi, M. Odersky (Eds.), Compiler Construction. XIV, 233 pages. 2007.

Vol. 4419: P.C. Diniz, E. Marques, K. Bertels, M.M. Fernandes, J.M.P. Cardoso (Eds.), Reconfigurable Computing: Architectures, Tools and Applications. XIV, 391 pages. 2007.

Vol. 4418: A. Gagalowicz, W. Philips (Eds.), Computer Vision/Computer Graphics Collaboration Techniques. XV, 620 pages. 2007.

Vol. 4416: A. Bemporad, A. Bicchi, G. Buttazzo (Eds.), Hybrid Systems: Computation and Control. XVII, 797 pages. 2007.

Vol. 4415: P. Lukowicz, L. Thiele, G. Tröster (Eds.), Architecture of Computing Systems - ARCS 2007. X, 297 pages. 2007.

Vol. 4414: S. Hochreiter, R. Wagner (Eds.), Bioinformatics Research and Development. XVI, 482 pages. 2007. (Sublibrary LNBI).

Vol. 4412: F. Stajano, H.J. Kim, J.-S. Chae, S.-D. Kim (Eds.), Ubiquitous Convergence Technology. XI, 302 pages. 2007.

Vol. 4411: R.H. Bordini, M. Dastani, J. Dix, A.E.F. Seghrouchni (Eds.), Programming Multi-Agent Systems. XIV, 249 pages. 2007. (Sublibrary LNAI).

Vol. 4410: A. Branco (Ed.), Anaphora: Analysis, Algorithms and Applications. X, 191 pages. 2007. (Sublibrary LNAI).

Vol. 4409: J.L. Fiadeiro, P.-Y. Schobbens (Eds.), Recent Trends in Algebraic Development Techniques. VII, 171 pages. 2007.

Vol. 4407: G. Puebla (Ed.), Logic-Based Program Synthesis and Transformation. VIII, 237 pages. 2007.

Vol. 4406: W. De Meuter (Ed.), Advances in Smalltalk. VII, 157 pages. 2007.

Vol. 4405: L. Padgham, F. Zambonelli (Eds.), Agent-Oriented Software Engineering VII. XII, 225 pages. 2007.

Vol. 4403: S. Obayashi, K. Deb, C. Poloni, T. Hiroyasu, T. Murata (Eds.), Evolutionary Multi-Criterion Optimization. XIX, 954 pages. 2007.

Vol. 4401: N. Guelfi, D. Buchs (Eds.), Rapid Integration of Software Engineering Techniques. IX, 177 pages. 2007.

Vol. 4400: J.F. Peters, A. Skowron, V.W. Marek, E. Orłowska, R. Słowiński, W. Ziarko (Eds.), Transactions on Rough Sets VII, Part II. X, 381 pages. 2007.

Vol. 4399: T. Kovacs, X. Llorà, K. Takadama, P.L. Lanzi, W. Stolzmann, S.W. Wilson (Eds.), Learning Classifier Systems. XII, 345 pages. 2007. (Sublibrary LNAI).

Vol. 4398: S. Marchand-Maillet, E. Bruno, A. Nürnberger, M. Detyniecki (Eds.), Adaptive Multimedia Retrieval: User, Context, and Feedback. XI, 269 pages. 2007.

Vol. 4397: C. Stephanidis, M. Pieper (Eds.), Universal Access in Ambient Intelligence Environments. XV, 467 pages. 2007.

Vol. 4396: J. García-Vidal, L. Cerdà-Alabern (Eds.), Wireless Systems and Mobility in Next Generation Internet. IX, 271 pages. 2007.

Vol. 4395: M. Daydé, J.M.L.M. Palma, Á.L.G.A. Coutinho, E. Pacitti, J.C. Lopes (Eds.), High Performance Computing for Computational Science - VEC-PAR 2006. XXIV, 721 pages. 2007.

Vol. 4394: A. Gelbukh (Ed.), Computational Linguistics and Intelligent Text Processing. XVI, 648 pages. 2007.

Vol. 4393: W. Thomas, P. Weil (Eds.), STACS 2007. XVIII, 708 pages. 2007.

Vol. 4392: S.P. Vadhan (Ed.), Theory of Cryptography. XI, 595 pages. 2007.

Vol. 4391: Y. Stylianou, M. Faundez-Zanuy, A. Esposito (Eds.), Progress in Nonlinear Speech Processing. XII, 269 pages. 2007.

Vol. 4390: S.O. Kuznetsov, S. Schmidt (Eds.), Formal Concept Analysis. X, 329 pages. 2007. (Sublibrary LNAI).

Vol. 4389: D. Weyns, H.V.D. Parunak, F. Michel (Eds.), Environments for Multi-Agent Systems III. X, 273 pages. 2007. (Sublibrary LNAI).

Vol. 4385: K. Coninx, K. Luyten, K.A. Schneider (Eds.), Task Models and Diagrams for Users Interface Design. XI, 355 pages. 2007.

Vol. 4384: T. Washio, K. Satoh, H. Takeda, A. Inokuchi (Eds.), New Frontiers in Artificial Intelligence. IX, 401 pages. 2007. (Sublibrary LNAI).

Vol. 4383: E. Bin, A. Ziv, S. Ur (Eds.), Hardware and Software, Verification and Testing. XII, 235 pages. 2007.

Vol. 4381: J. Akiyama, W.Y.C. Chen, M. Kano, X. Li, Q. Yu (Eds.), Discrete Geometry, Combinatorics and Graph Theory. XI, 289 pages. 2007.

Vol. 4380: S. Spaccapietra, P. Atzeni, F. Fages, M.-S. Hacid, M. Kifer, J. Mylopoulos, B. Pernici, P. Shvaiko, J. Trujillo, I. Zaihrayeu (Eds.), Journal on Data Semantics VIII. XV, 219 pages. 2007.

Vol. 4379: M. Südholt, C. Consel (Eds.), Object-Oriented Technology. VIII, 157 pages. 2007.

Vol. 4378: I. Virbitskaite, A. Voronkov (Eds.), Perspectives of Systems Informatics. XIV, 496 pages. 2007.

Vol. 4377: M. Abe (Ed.), Topics in Cryptology – CT-RSA 2007. XI, 403 pages. 2006.

Vol. 4376: E. Frachtenberg, U. Schwiegelshohn (Eds.), Job Scheduling Strategies for Parallel Processing. VII, 257 pages. 2007.